NUTRITION AND DIET RESEARCH PROGRESS

HUMAN MILK

NUTRITIONAL CONTENT AND ROLE IN HEALTH AND DISEASE

NUTRITION AND
DIET RESEARCH PROGRESS

Additional books and e-books in this series can be found on Nova's website under the Series tab.

NUTRITION AND DIET RESEARCH PROGRESS

HUMAN MILK

NUTRITIONAL CONTENT AND ROLE IN HEALTH AND DISEASE

JOHN I. COLE
EDITOR

Copyright © 2021 by Nova Science Publishers, Inc.

All rights reserved. No part of this book may be reproduced, stored in a retrieval system or transmitted in any form or by any means: electronic, electrostatic, magnetic, tape, mechanical photocopying, recording or otherwise without the written permission of the Publisher.

We have partnered with Copyright Clearance Center to make it easy for you to obtain permissions to reuse content from this publication. Simply navigate to this publication's page on Nova's website and locate the "Get Permission" button below the title description. This button is linked directly to the title's permission page on copyright.com. Alternatively, you can visit copyright.com and search by title, ISBN, or ISSN.

For further questions about using the service on copyright.com, please contact:
Copyright Clearance Center
Phone: +1-(978) 750-8400   Fax: +1-(978) 750-4470   E-mail: info@copyright.com.

NOTICE TO THE READER

The Publisher has taken reasonable care in the preparation of this book, but makes no expressed or implied warranty of any kind and assumes no responsibility for any errors or omissions. No liability is assumed for incidental or consequential damages in connection with or arising out of information contained in this book. The Publisher shall not be liable for any special, consequential, or exemplary damages resulting, in whole or in part, from the readers' use of, or reliance upon, this material. Any parts of this book based on government reports are so indicated and copyright is claimed for those parts to the extent applicable to compilations of such works.

Independent verification should be sought for any data, advice or recommendations contained in this book. In addition, no responsibility is assumed by the Publisher for any injury and/or damage to persons or property arising from any methods, products, instructions, ideas or otherwise contained in this publication.

This publication is designed to provide accurate and authoritative information with regard to the subject matter covered herein. It is sold with the clear understanding that the Publisher is not engaged in rendering legal or any other professional services. If legal or any other expert assistance is required, the services of a competent person should be sought. FROM A DECLARATION OF PARTICIPANTS JOINTLY ADOPTED BY A COMMITTEE OF THE AMERICAN BAR ASSOCIATION AND A COMMITTEE OF PUBLISHERS.

Additional color graphics may be available in the e-book version of this book.

Library of Congress Cataloging-in-Publication Data

ISBN: 978-1-53619-713-6

Published by Nova Science Publishers, Inc. † New York

CONTENTS

Preface vii

Chapter 1 Impressive Antimicrobial Potential of
Human Milk: A Blessing for Nascent Life 1
*Nivedita Sharma, Bindu Devi
and Moksharthi Sharma*

Chapter 2 Fatty Acid Supply of Human Milk
and Its Possible Health Effects 69
Éva Szabó

Chapter 3 Life-Course Health Impacts of the
Nutritional Content of Human Milk in
Different Steps of Lactation 109
*Parisa Iravani, Motahar Heidari-Beni
and Roya Kelishadi*

Index 127

PREFACE

Mother's milk provides a wide variety of health benefits for infants, which is why breast feeding is highly recommended all over the world. This monograph provides details about the nutritional and bioactive properties of human milk, explaining the significance of this natural source of food. Chapter One describes how lactic acid bacteria present in human milk plays a critical role in establishing an immunocompetent microbiome in newborns, inducing multifarious health-promoting activities required to successfully initiate the life process. Chapter Two describes how the fatty acid profile of breastmilk varies according to the needs of the infant, particularly depending on the gestational age of the newborn, but also as a result of the mother's diet. Chapter Three summarizes the current literature on the composition of human milk and its life-course functional effects on health outcomes.

Chapter 1 - Human milk is considered a lifeline for the neonates; which is why breast feeding is highly recommended and promoted all over the world. Colostrum, the very first mother after child birth, is medically proven to be essential for the wellbeing of the new born. Mother's milk has endless protective benefits for babies besides providing basic nutrition and is, therefore, considered the best food for infants. Till date, no alternative has been found that can replace the mother's milk. One of the most salient benefits of human milk is its broad range antimicrobial properties, especially against challenging human pathogens that otherwise can be life threatening to the little one. Human milk samples collected from new mothers admitted in different maternity wards of the hospitals/households when assessed for their antagonistic potential have been found to be enriched with lactic acid bacteria (LAB). Upon further investigation, their antimicrobial properties have been found to be directly related to the lactic acid bacteria present in breast milk. Many of these LAB were identified using 16S-rRNA genotyping and their phylogenies have also been traced. The LAB isolates on testing are found to exertimpressive inhibition against human neonatal pathogens selected as test strains. The wide zones of inhibitions strongly affirm their broad range antimicrobial activity against a broad range Gram positive as well as Gram negative pathogens. Further, these LAB when evaluated have revealed strong probiotic attributes proving their critical role in establishing a much-desired immunocompetent microbiome in the new born to induce multifarious health promoting activities required to initiate the life process successfully.

Chapter 2 - Human milk is the optimal choice for infant nutrition, with exclusive breastfeeding for the first six months of life and introduction of complementary feeding should also start along with continued breastfeeding. Long chain polyunsaturated fatty acids (LCPUFAs) play an important role in building up membranes and they are also precursors of certain second messengers. The two most important LCPUFAs, the n-3 docosahexaenoic acid (DHA) and n-6 arachidonic acid (AA) are the predominant fatty acids in human brain and are key factors in the visual- and neurodevelopment during the third trimester and in the first months of life. For breastfed infants the exclusive source of these fatty acids is human milk which can be influenced by maternal diet. Several studies have found a difference between breastmilk of mothers who gave birth to term newborn and breastmilk of mothers who gave birth of preterm infants. Preterm neonates have only limited body stores of LCPUFAs but their dietary requirement for these fatty acids is increased due to their rapidly developing tissues. Therefore, the fatty acid composition of breast milk of mothers who gave birth to preterm or term, mature babies is different in order to cover their different dietary needs. The fatty acid composition of breastmilk is influenced not only by the gestational age at birth, but also by the time of sampling. Several studies have found that fatty acid composition of colostrum is different from that of mature milk. These results suggest that fatty acid profile of breastmilk varies throughout lactation adapting to the nutritional needs of the developing infant. The main sources of fatty acids in human milk are maternal diet as well as maternal stores. Therefore, some of the differences found in the fatty acid composition of milk samples can be attributed to different eating habits in different countries. This chapter gives an overview of the fatty acid composition of breastmilk in mothers who gave birth to newborn of different gestational ages, as well as during lactation or as a result of diet.

Chapter 3 - Breastfeeding has numerous health benefits both for the infants and the nursing mothers. In addition to its short-term benefits, it is documented that breastfeeding has preventative roles against chronic non-communicable diseases including hypertension, obesity, diabetes, hyperlipidemia, and cardiovascular diseases in adulthood. These health benefits of human milk are correlated with its nutritional and bioactive components including antimicrobial substances, anti-inflammatory components and hormones, which modulate the body metabolism and composition. Breastfeeding duration is also one of the factors that might determine the amount of biological effects of milk. The composition of milk changes constantly during lactation to provide nutritional necessities to protect infant against potential harmful pathogens. In addition, gestational length, maternal diseases and nutrition influence the composition of human milk. The protein content of human milk gradually reduces from the second to the sixth or seventh month of lactation and stabilizes at the final step. Protein concentration is associated with body weight gain of infant and immune protection, as well as increase in nutrient digestion and availability of micronutrient. The fat content increases during lactation and would affect neurological development and cognitive outcome of infant. Human milk carbohydrates impact on the appetite regulation and body composition, which would protect infant against obesity. Understanding the human milk composition over time and its health benefits can be important for primordial prevention of non-communicable diseases. This chapter aims to summarize the current literature on the composition of human milk and its life-course functional effects on health outcomes.

In: Human Milk　　　　　　　　　　　ISBN: 978-1-53619-713-6
Editor: John I. Cole　　　　　　　　© 2021 Nova Science Publishers, Inc.

Chapter 1

IMPRESSIVE ANTIMICROBIAL POTENTIAL OF HUMAN MILK: A BLESSING FOR NASCENT LIFE

Nivedita Sharma[1,*], *Bindu Devi*[1]
and Moksharthi Sharma[2]

[1]Microbiology Research Laboratory, Department of Basic Sciences,
Dr. YS Parmar University of Horticulture and Forestry,
Nauni, Solan, H.P., India
[2]Sikkim Manipal Institute of Medical Sciences, Gangtok, Sikkim, India

ABSTRACT

Human milk is considered a lifeline for the neonates; which is why breast feeding is highly recommended and promoted all over the world. Colostrum, the very first mother after child birth, is medically proven to be essential for the wellbeing of the new born. Mother's milk has endless protective benefits for babies besides providing basic nutrition and is, therefore, considered the best food for infants. Till date, no alternative has been found that can replace the mother's milk. One of the most salient benefits of human milk is its broad range antimicrobial properties,

---

[*] Corresponding Author's E-mail: niveditashaarma@yahoo.co.in.

especially against challenging human pathogens that otherwise can be life threatening to the little one. Human milk samples collected from new mothers admitted in different maternity wards of the hospitals/households when assessed for their antagonistic potential have been found to be enriched with lactic acid bacteria (LAB). Upon further investigation, their antimicrobial properties have been found to be directly related to the lactic acid bacteria present in breast milk. Many of these LAB were identified using 16S-rRNA genotyping and their phylogenies have also been traced. The LAB isolates on testing are found to exert impressive inhibition against human neonatal pathogens selected as test strains. The wide zones of inhibitions strongly affirm their broad range antimicrobial activity against a broad range Gram positive as well as Gram negative pathogens. Further, these LAB when evaluated have revealed strong probiotic attributes proving their critical role in establishing a much-desired immunocompetent microbiome in the new born to induce multifarious health promoting activities required to initiate the life process successfully.

Keywords: colostrum, mature milk, antimicrobial factors, immunoglobulins, human pathogens

INTRODUCTION

"Mother is God's most precious gift on earth
And Her milk is God's nectar for a baby"

Human milk and colostrum are very complex biological fluids characterized by the genes and the environment of the producing mother to provide nutritional, protective and developmental functions to suit all the needs of the developing neonate in an age-adapted manner. Both human milk and colostrum contain all the essential nutrients required for crucial immunological protection and are considered as a wholesome diet for the neonate during the initial phase of its life (Jost et al. 2015; Italianer et al. 2020). The medical dictionary defines 'Colostrum' as the first milk secreted at the time of parturition. Being richer in lactalbumin and lactoprotein than the milk secreted later, Colostrum is rich in antibodies that confe acquired immunity on the newborn, also called "foremilk" (Samuel et al. 2020; Godhia and Patel, 2013). Colostrum has a significant role to play in the immune system of every mammal and lasts for 2-4 days after lactation. Human colostrum (HC) is especially rich in antimicrobial peptides (lactoferrin and lactoperoxidase), immunoglobulin (Ig), bioactive molecule, including growth factors that are important for nutrition, growth and development of newborn infants and also for acquired immunity (Figure 1 and 2) (Drozdowski et al. 2010). This "early" milk has a nutrient profile and immunological composition significantly different from 'mature' milk in terms of its exceptionally strong antipathogenic potential. Human milk (Figure 1 and 2) contains macronutrients like proteins, carbohydrates, oligosaccharides, fats and micronutrients like vitamins and minerals, growth factors, anti-microbial compounds and immune regulating constituents either not present in milk or present substantially in a lower concentration (Samuel et al. 2020; Golinelli et al. 2011). A neonate is exposed to a myriad of pathogens as soon as h/she exits the sterile and safe womb of the mother with a relatively immature immune system which is not well equipped to combat such a situation. Thus, the baby is rendered susceptible to a wide variety of infections such as sepsis, gastrointestinal infections, respiratory tract infections, meningitis, etc. which are amongst the leading causes of morbidity and mortality in new-borns and infants all around the globe. In order to help the baby withstand these infectious agents until its immune system develops, certain compounds of immunity are provided through the breast milk. Colostrum and breast milk, besides being nutritionally rich, are also plentiful in a variety of bioactive compounds which protect the baby; colostrum is, therefore, sometimes even referred to as a natural vaccine. Neonatal gut microbiota establishment represents a vital stage for gut maturation, metabolic and immunologic programming and consequently short and long-term health status. Human milk beneficially influences this process thanks to its dynamic profile of age-adapted nutrients and bioactive components and by providing commensal maternal bacteria to the neonatal gut. These include *Lactobacillus* spp., also as obligate anaerobes like *Bifidobacterium* spp., which can originate from the maternal gut via an enteromammary pathway as a completely unique sort of mother–neonate communication. This makes human milk the indisputable gold standard for early nutrition which no formula milk (till date) can compete with (Spatz et al. 2021; Walker, 2010). Compared with formula feeding, breastfeeding has long been associated with lower infant morbidity and mortality, as well as with reduced incidence and severity of infections and metabolic and immune-related diseases (Turin and Ochoa, 2014). The well-recognized benefits of human milk on health status, even beyond the period of breastfeeding, are not only attributable to its adapted macro- and micronutrient composition, i.e., carbohydrates, lipids, proteins and minerals and vitamins, respectively, from colostrum to late lactation, but to the wide range of bioactive components that may act individually and/or synergistically (Samuel et al. 2020; Isaacs, 2005). These components include digestive and antioxidative enzymes (e.g., amylase and lipase, lactoperoxidase and SOD, respectively), growth/trophic factors (e.g., transforming growth factor-beta, epidermal protein, insulin-like growth factor), antimicrobial compounds (e.g., lactoferrin), and compounds that provide passive immunity (e.g., lysozyme, secretory IgA, cytokines and leukocytes) to the structurally and functionally immature gut that lacks immune competence (Quitadamo et al. 2021; Wagner et al. 2008).

(Source: https://en.wikipedia.org/wiki/Breast_milk#/media/File:Human_Breastmilk_-_Foremilk_and_Hindmilk.png)

Figure 1. Two 25-milliliter samples of human breast milk. The left-hand sample is the first milk produced by the mother while the right-hand sample has been produced later.

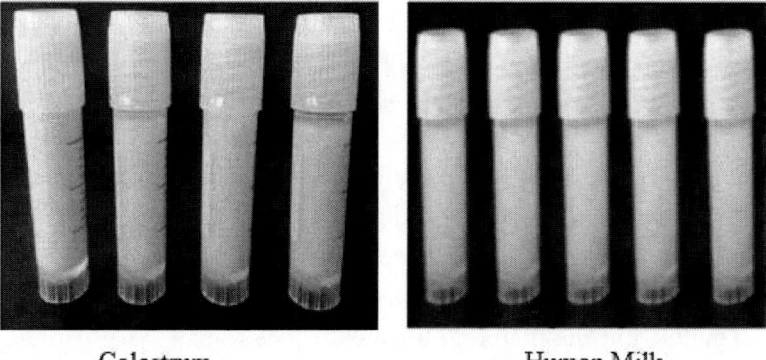

Colostrum   Human Milk

Figure 2. Human milk and colostrum samples collected from maternity wards of Kamala Nehru Hospital Shimla, H.P. (India).

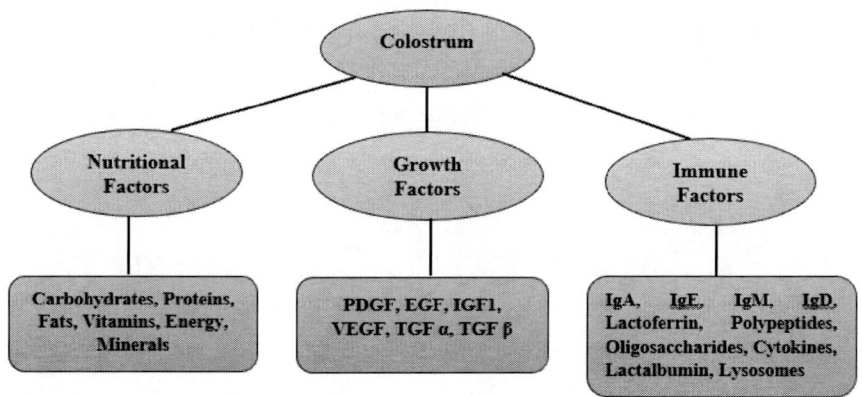

(Source: Godhia and Patel, 2013)

Figure 3. Composition of Colostrum.

Human milk used to be traditionally considered sterile; however, recent studies have shown that it makes available a continuous supply of commensal, mutualistic and/or potentially probiotic bacteria to the infant gut. Recently, human milk has been recognized as a continuous source of viable commensal maternal bacteria able to colonize the neonatal gut, in dominantly lactic acid bacteria. Culture-dependent and culture-independent techniques have revealed the dominance of Staphylococci, Streptococci, Lactic acid bacteria and *Bifidobacteria* in this biological fluid and their role in the colonization of the infant gut (Jost et al. 2013). Different studies suggest that some bacteria present in the maternal gut could reach the mammary gland during late pregnancy and lactation through a mechanism involving gut monocytes. These bacteria could protect the infant against infections by producing different antimicrobials, such as lactic acid, acetic acid, hydrogen peroxide, carbon dioxide and bacteriocins (Handa and Sharma, 2016) and contribute to the maturation of the immune system, among other functions. It has been shown that some human milk bacteria may originate from the maternal gut via an enteromammary pathway and prepare the neonatal immune system for the maternal environment (Jimenez et al. 2008). Human milk also harbors structurally diverse, nondigestible oligosaccharides (human milk oligosaccharides [HMOs]) that can enhance the growth of specific gut bacteria (Bode, 2009; Luna et al. 2020). The aim of this study is to present current knowledge regarding evidence on the importance of immunocompetent microbiota established through breastfeeding to prevent from infections and the functionality of selected human milk ingredients in infants.

COMPOSITION OF COLOSTRUM

Colostrum supports the human organism in two main ways. Firstly, its multiple immune factors and natural antibiotics provide strong support for the immune system; secondly, its numerous growth factors offer a broad-spectrum boost to the organism thus encouraging optimum health and healing. The immunoglobulin, growth factors and antibodies play an important role in the prevention of infection i.e., passive immunity (Spatz et al. 2021; Golinelli et al. 2011). The vital nutrients help in tissue development, growth and energy. The growth factors present in the colostrum provide a novel treatment option for gastrointestinal conditions. The Table 1 and Figure 3 indicate nutrient content, immunological factors and growth factors of colostrum (Golinelli et al. 2014; Quintadamo et al. 2021). Colostrum contains the growth factors that help build lean muscle, including insulin-like growth factors (IGF-I & IGF-II) and growth hormone (GH). IGF-I, which is found naturally in colostrum, is the only natural hormone capable of promoting muscle growth by itself. The growth factors in colostrum "shift fuel utilization from carbohydrate to fat (Golinelli et al. 2014). It means that body will burn more fat, including the fat made from carbohydrate and protein consumption, producing fuel more efficiently (Bhagwe-Parab et al. 2020). The IGF-1 in colostrum increases the uptake of blood glucose and facilitates the transport of glucose to the muscles, which keeps energy levels up.

Together with growth hormones, IGF-1 also slows the rate of protein breakdown (catabolism) that occurs after a vigorous workout. It speeds up protein synthesis, which results in lean muscle mass without an increase in the amount of stored fat. Colostrum improves the assimilation of nutrients, which leads to improved energy levels and performance. Also, the immune factors in colostrum help newborns to minimize their susceptibility to infections (Godhia and Patel, 2013; Bhagwe-Parab et al. 2020).

Table 1. Nutritional, immune and growth factors inhuman colostrum

Nutritional	**Factor**	**Human Colostrum**
	Energy(kcal)	58
	Protein(g)	3.7
	lactose(g)	5.3
	Fat(g)	2.9
Immune (mg/mL)	Lactoferrin	700
	IgA	17.35
	IgG	0.43
	IgG2	-
	IgM	1.59
Growth	Epidermal growth factor (EGF)	200mcg/L
	Transforming growth factor (TGF α)	2.2–7.2 mcg/L
	TGF β	20–40 mg/L
	Insulin like growth factor (IGF)	18 mg/L
	Vascular endothelial growth factor	75 mcg/L
	Growth hormone (GH)	41ng/L

(Source: Godhia and Patel, 2013)

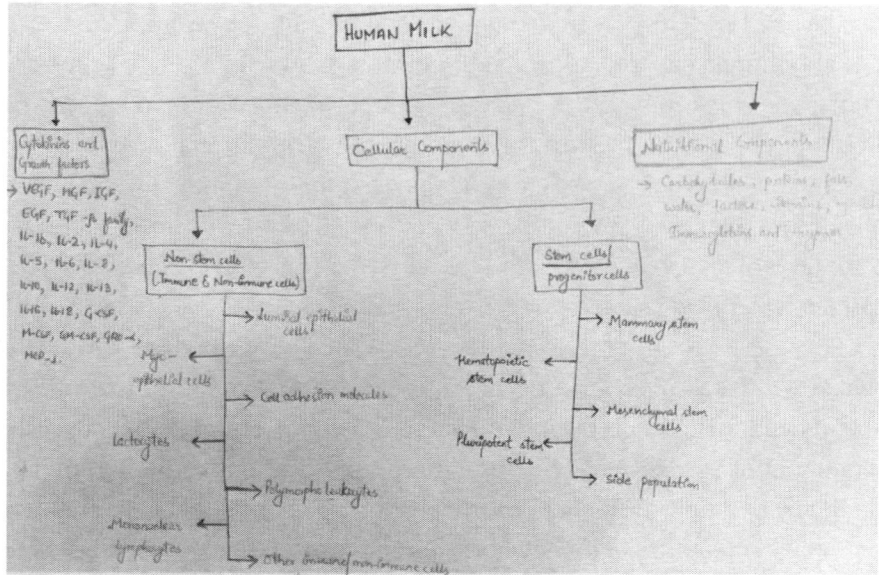

(Source: Kaingade et al. 2016)

Figure 4. Composition of human milk.

COMPOSITION OF HUMAN MILK

Immunoglobulins

Immunoglobins are the major constituent of the protein content of milk. They are known to carry the immunological memory of the mother to the baby thereby providing passive immunity against infections that the mother has encountered during her lifetime (Figure 4) (Mantis et al. 2011).

Secretory IgA

Breast milk contains immunoglobulins with antibody activity against several micro-organisms. Hitherto, these antibodies have been supposed to be of minor importance since they are not absorbed by the gut of the human infant but this assumption became questionable when it was discovered that IgA was the predominant immunoglobulin in milk in contrast to serum and that the milk IgA was antigenically different from serum IgA. It is the most abundant and the most important immunoglobulin in human milk. Milk IgA forms part of a local defence system consisting of locally produced 'secretory IgA' making up the dominating immunoglobulin in most secretions (Golinelli et al. 2014). It has been suggested that this secretory immunoglobulin effects the antimicrobial protection of mucous membranes. These facts make it likely that the function of milk antibodies will be locally acting, primarily in the gastrointestinal tract. Antibody activity against other micro-organisms like polio virus, streptococci and pneumococci has also been shown in the secretory IgA fraction of human milk. The secretory IgA molecule is more resistant to pH changes and proteolytic enzymes than is serum IgA or other immunoglobulins which may enable the secretory IgA antibodies of milk to function in the variable milieu of the gut. Some preliminary experiments have indicated, however, that a nucleotide dependent reductase isolated from liver can split secretory IgA (Kaingade et al. 2016). IgA is mostly attributed to protection of mucosal membranes aimed mostly against enteric and respiratory pathogens. The suckling baby receives about 0.25-0.5g of IgA through breast milk every day. It is resistant to intestinal enzymes (Hansen and Soderstrom, 1981).

Enteric Protection

The enteric protection is effected by preventing binding and penetration of pathogens into gut mucosa, thereby, preventing infections and providing mucosal immunization. The access to receptors present on the epithetlium is blocked and the pathogens are trapped in mucus and removed by peristalsis (Mantis et al. 2011). sIgA also selectively binds to cells in Peyer's patches and is transported to the lymphoid tissue.

Table 2. Other components responsible for antimicrobial properties of breast milk

Components	Action
True secretory component	Inhibition of epithelial adhesion *E. coli*, *Salmonella typhimurium*, *Pneumococcus*, etc. Inhibition of toxins like Closteridium difficle toxin A (Mantis et al. 2011)
Lactoferrin	Conversion to potent cationic peptide Lactoferricin Disintegration of external lipopolysaccharide membrane in gram negative bacteria (Mantis et al. 2011) Bacteriostatic action due to iron chelation (Mantis et al. 2011;Goldman et al. 2020)
Lysozymes	Antiviral activity (Mantis et al. 2011) Cause degradation of outer cell wall of gram-positive bacteria by hydrolysis (Mantis et al. 2011)
α-lactalbumin	Partially unfolded α-lactalbumin + oleic acid ⟶ HAMLET (**Human** α-lactalbumin made lethal to human cells) kills different types of tumor cells (Goldman et al. 2020) Antimicrobial activity against *E. coli, K. pneumoniae, Staphylococcus, Streptococcus* and *Closteridium albicans*
κ-casein	Glycoprotein charged sialic acid residue Soluble receptor analogue of epithelial cell surfaces Inhibition of *Helicobacter pylori* adhesion (Goldman et al. 2020)
Lactoperoxidase	Thiocynate(saliva) ⟶ Hypothiocynate ⟶ bactericidal to both gram negative and gram positivebcteria (Mantis et al. 2011)
Lactadherin	Binds to vitamin B12 ⟶ Bacteriostatic (Mantis et al. 2011)
Macrophage Migration Inhibitory Factory (MIF)	Proinflammatory cytokines Bactericidal for *Mycobacterium tuberculosis* (Goldman et al. 2020)
Complement	All compounds can be found but in low concentration C3 degaradtion products supplement pathogen phagocytosis in GIT and respiratory tract (Goldman et al. 2020)
Mucin	Blocks E. coli adhesion to epithelial cells (Goldman et al. 2020) in stomach and small intestine (Mantis et al. 2011)
Lipids	True fatty acids: - Antiprotzoal especially against Giardia and *Entamoeba histolytica* - Virolyic Monoglycerides: - Virolytic (Mantis et al. 2011) - Hydrolysis of certain bacteria (Goldman et al. 2020)
Carbohydrates	Growth of Bifidobacterium and Lactobacilli Inhibition of pathpgenic Growth (Mantis et al. 2011) Direct antibacterial effects against enteric bacteria and epithelial binding Absorbed and excreted in to urinary tract- Protection against UTI's

When foreign antigens are encountered and presented to T cells, these secrete cytokines and local cytotoxic T cells (Goldman et al. 2020) and consequently provide long term protection by maintaining healthy gut microbiota. Pathogens like *E. coli*, *H. pylori*, *Shigella* sp., *Salmonella* sp., *Vibrio cholera*, *Cytomegalo* virus, Rotavirus, Poliovirus, *Giardia lamblia* (Palmeira P and Carneiro-Sampaio M, 1992).

Respiratory Epithelium

It acts in a similar manner to the action in protection of gut epithelium by coating and providing local first line of defence to respiratory epithelium against pathogens such as *Kleibsella pneumoniae*, *Streptococcus pneumoniae*, *Haemophilus influenzae*, Adenovirus, Respiratory syncytial virus, *Candida albicans*. The concentration of sIgA declines over the course of breast feeding but the availability to the baby is maintained due to increased intake of the milk (Mantis et al. 2011).

Secretory IgM

It is the second most ample immunoglobulin found in human milk. Its function is similar to that of sIgA and works in adjunct with it to provide local mucosal immunity (Mantis et al. 2011).

IgG

It is found in very low concentration (0.1mg/ml) and has the potential to activate the compliment system (Mantis et al. 2011). Its most abundant subclass is IgG4 which is present in relatively higher proportions in milk than in human serum (Goldman et al. 2020).

Proteins

The protein content of human milk decreases rapidly during the primary month of lactation and declines at a lot slower rate later.. Most proteins are synthesized by the mamma, with a couple of possible exceptions, like albumin (which appears from the maternal circulation). Milk proteins are often classified into 3 groups: mucins, caseins and whey proteins. Mucins, also referred to as milk fat globule membrane proteins, surround the lipid globules in milk and comprise only a little percentage of the entire protein content of human milk. Since the fat content of human milk doesn't vary during the course of lactation, the milk mucin concentration is presumably constant, although little information is available regarding this subject. The contents of casein and whey proteins, however, change profoundly early in lactation; the concentration of whey proteins is extremely high, whereas casein is virtually undetectable during the primary days of lactation. Subsequently, casein synthesis within the mammary gland and milk casein concentrations increase, whereas the concentration of total whey proteins decreases, partially due to an increased volume of milk being produced. As a consequence, there's no "fixed" ratio of whey to casein in human milk since it varies throughout lactation. The frequently cited ratio of 60:40 is an approximation of the ratio during the traditional course of lactation, but it does vary from 80:20 in early lactation to 50:50 in late lactation. Since the amino acid compositions of caseins and whey proteins differ, the aminoalkanoic acid content of human milk also varies during lactation. The true protein content of breast milk, determined as described above, is 14–16 g/L during early lactation, 8–10 g/L at 3–4 month of lactation, and 7–8 g/L at 6 month and later (Conte-Junior et al. 2006). It has been argued that these protein concentrations and consequently, the corresponding intakes don't accurately reflect the utilizable amounts of amino acids provided to breastfed infants. This argument is predicated on the observation that intact breast-milk proteins are often found within the stool of breastfed infants. Thus, they're incompletely digested and don't represent utilizable amino acids. It has, therefore, been suggested that such proteins [e.g,, lactoferrin and secretory immunoglobulin A (sIgA)] should be subtracted from the protein concentration of breast milk to derive a real digestible protein content (Hamosh, 2001). However, it's not correct to assume that these proteins are completely indigestible; albeit they need properties that make them more immune to proteolytic enzymes than are most other proteins. It is only a minor fraction that escapes digestion; for example, it has been estimated that 6–10% of lactoferrin is not digested by breastfed infants. The quantitatively most vital of the relatively indigestible human milk proteins are lactoferrin and sIgA. The total concentration of these 2 proteins in mature milk (> 30 d of lactation) is 2 g/L. Assuming that 6–10% is undigested, there is a potential loss of 0.12–0.2 g/L (or 1.2–2.0% of total protein intake), which may be within the margin of error for the analysis used. Thus, whereas undigested, biologically active proteins may have physiologic significance for breastfed infants, the effect of the loss of those amino acids on the nutrition of infants could also be insignificant (Chang et al. 2019; Golinelli et al. 2014).

CHANGES IN THE COMPOSITION OF PROTEINS IN HUMAN MILK DURING THE FIRST MONTHS OF LACTATION

Colostrum

From the second trimester of pregnancy, a mother's mammary gland is sufficiently active to supply breast milk. Throughout the milk production period, constant variations in milk composition, and differences are observed from one woman to a different and in between samples taken from an equivalent woman during an equivalent day and even during an equivalent feeding session [Chang et al. 2019; Mataix and Hernandez, 2002]. Even during these changes, human milk always maintains a balanced chemical composition regarding the demand for the required nutrients required for rapid climb and maturation of tissue, and at an equivalent time, for the organs involved within the regulation of endogenous metabolism. Still, given this massive variation, it's vital to understand how the milk samples are collected. The idea is to gather a sample of all the milk produced within 24 hours and at different times of lactation [Spartz et al. 2021; Conte-Junior et al. 2006]. During the last trimester of pregnancy, the mamma accumulates within the lumen of the alveoli pre-colostrum, chiefly composed of plasma exudate, cells, immunoglobulins, lactoferrin, serum albumin, ions like sodium and chlorine, and a little amount of lactose. On an average, during the primary 7 days after delivery, colostrum is produced. The primary product of the of lactic nurse secretion, it's a yellow, high denisty, low volume fluid. During the first few days, 220ml of colostrums per dose is produced, which is sufficient to satisfy the necessities of a newborn. Colostrum incorporates a lower energy content, lactose, lipids, glucose, urea, water soluble vitamins and nucleotides than mature milk, however, it's a greater protein content, sialic acid, lipid-soluble vitamins E, A, K and carotenes than mature milk [Gazzolo et al. 2004]. The concentrations of minerals like zinc, sodium, iron, sulfur, selenium, manganese and potassium also are higher in colostrum. The ratio of whey proteins/casein ratio is 80/20 in colostrum, whereas in mature milk, it is 60/40 and 50/50 in the late lactation [Dupont, 2003; Luna et al. 2021]. Likewise, the concentration of free amino acids varies in comparison to colostrum, transitional milk and mature milk. the quantity of protein decreases rapidly during the primary month then stabilizes for a time, but decreases slowly over the lactation. Colostrum features a high content in immunoglobulins particularly IgA, lactoferrin, cells (lymphocytes and macrophages), oligosaccharides, cytokines and other defense factors that protect newborns from microorganisms of the environment and promote the maturation of the system (Table 3). This fluid is tailored to the precise needs of the neonates since their immature kidneys cannot filter large volumes of liquids ADDITIONALLY it also facilitates the elimination of meconium, preventing hyperbilirubinemia of the newborn [Marnila and Korhonen, 2003]. Colostrum contains intestinal enzymes that help in digestion (lactase and other intestinal enzymes aren't produced within the newborn). The high concentration of immunoglobulins allows the endothelium of the alimentary canal to be covered preventing the adhesion of pathogens. Furthermore, colostrum has antioxidants and quinone that protect the alimentary canal from oxidative damage, it's also rich in growth factors that stimulate the maturation of both the digestive and immune systems [Golinelli et al. 2014; Spatz et al. 2021].

Table 3. Difference in carbohydrates, vitamins and micronutrients content in human colostrum and mature milk

Components	Human Colostrum	Human Milk
Water, g	-	88
Protein, g	2.7	1.2
Lactose, g	5.3	6.8
Casein: α-Lactalbumin ratio	-	1:2
Fat, g	2.9	3.8
Linoleic acid	-	8.3% of fat
Potassium, mg	55	55
Sodium, mg	92	15
Calcium, mg	31	33
Magnesium, mg	4	4
Chloride, mg	117	43
Phosphorous, mg	14	15
Iron, mg	0.09	0.15
Vit D, ug	-	0.03
Vit A, ug	89	53
Riboflavin, ug	30	43
Nicotinic acid, ug	75	172
Ascorbic acid, mg	4.4	4.3
Thiamine, ug	15	16

(Source: Gollinelli et al. 2014)

Transitional/Mature Milk

The term transitional milk, or intermediary, refers to milk that's produced after the colostrum, on the average between 7 and 21 days postpartum. During this period there's anincrement in milk production, which follows a sequential increase until it reaches a volume of 600-700ml per day, between the 15th to 30th day of lactation. This milk has an intermediate composition and varies from day to day, until it reaches the composition of mature milk. Changes in milk composition occur more slowly during this period than within the period immediately after birth [Martínez-García, 2002; Spatz, 2020]. After this period, what is called mature milk is produced and secreted, on a mean, after 21 days postpartum. This composition also shows variability, but less than that observed during early lactation, the intermediate concentrations of fat and lactose increases, whereas the concentrations of proteins, growth factors, immune factors, particularly IgA, and minerals decrease [Rivero-Urgell et al. 2005; Takako et al. 2020]. Mature milk contains a spread of nutritive and nonnutritive components. The average volume of milk produced by a mature woman is 700-900 ml per day during the primary 6 months postpartum, and if the mother has twins, enough volume for every baby is produced. Moreover, lactation progresses to a phase of colostrum before merging into milk secretion. The entire protein content of human milk undergoes longitudinal variation, however, unlike the fat little transverse variation. The entire protein content is higher in colostrum (approximately 15.8 gL-1) and reduces during lactation (about 9.0 gL-1 in mature milk). The protein content ingested by the infant, however, doesn't vary much during lactation, since the quantity of ingested colostrum produced is smaller than the quantity of mature milk, which compensates for variations in protein concentration during lactation [Lönnerdal and Atkinson, 1995; Spatz, 2020]. The concentration of whey protein, which is additionally higher in colostrum is reduced with the passage of time in lactation. These changes end in a casein/whey protein ratio of roughly 10:90 within the first few days of lactation to 45:55 in mature milk. the ratio of casein/whey protein in cow's milk is about 80:20, this ratio is sort of different from that contained in human milk. A change in milk composition during lactation is most pronounced during the primary weeks of lactation (Takako et al. 2020; Golinelli et al. 2014).

Biological Effects of Human Milk

The proteins present in the serum of colostrum and breastmilk have several nutritional and physiological functions [Chang et al. 2019;

Korhonen et al. 1998], some examples of the functions of these proteins are mentioned in table 4 in order to understand the large role played by these proteins, and their importance for the newborn. In addition to the proteins mentioned above, there exist other serum proteins, whose function has already been established and others whose functions are not yet clear [Ma et al. 2020; Golinelli et al. 2014].

κ-Casein

κ-Casein, a minor casein subunit in human milk, is a glycoprotein with charged sialic acid residues. The heavily glycosylated α-casein molecule has been shown to inhibit the adhesion of *Helicobacter pylori* to human gastric mucosa. "*H. pylori* infection has been shown to occur in increasingly newborns, but breastfeeding seems to provide some protection. It is likely that the carbohydrate component of α-casein is responsible for this activity because sIgA, which is also glycosylated, exhibits similar activity, and both proteins lose their activity when deglycosylated" (Golinelli et al. 2014). κ-Casein has been shown to forestall the attachment of bacteria to the mucosal lining by acting as a receptor analogue. Oligosaccharide structures on the glycans of those glycoproteins act as decoys for similar surface-exposed carbohydrate structures on the gastric mucosa, thereby inhibiting adhesion. Lactoferrin has been shown to inhibit the growth of *H. pylori in vitro* and it is thus possible for lactoferrin, α-casein and sIgA work together to limit the growth, proliferation and adhesion of this pathogen (Golinelli et al. 2011; Ma et al. 2020).

Haptocorrin

Only a small percentage of the vitamin B-12–binding capacity of haptocorrin is occupied in human milk, leaving it during??? a very unsaturated form. It has been suggested that vitamin B12–binding protein (haptocorrin) inhibits bacterial growth by tightly binding and withholding the vitamin from the bacteria (Ma et al. 2020; Adkins and Lönnerdal, 2002). The structure and activity of haptocorrin was maintained after *in vitro* digestion with pepsin and pancreatin, indicating that haptocorrin may resist digestion within the gut. Whether this is often the inhibiting mechanism, how broad its antimicrobial activity is and whether haptocorrin quantitatively contributes to the defense against infection in breastfed infants remain to be explored. Recent studies *in vitro* show that both apo- and holo-haptocorrin can inhibit the expansion of EPEC, but the mechanism of this vitamin B-12–independent activity isn't yet known (Lonnerdal, 2003; Luna et al. 2020).

α-Lactalbumin

Few studies have focused on the potential antimicrobial activity of α - lactalbumin. However, 3 polypeptide fragments from α-lactalbumin were recently found to possess antimicrobial activity against *E. coli, Klebsiella pneumoniae, Staphylococcus aureus, Staphylococcus epidermis, Streptococci* and *C. albicans*. These peptides were generated after exposure to proteases known to be present within the alimentary canal. This may explain our finding of an inhibitory effect of α-lactalbumin supplemented infant formula on EPEC-induced diarrhea in infant rhesus monkeys. The α-lactalbumin is a major protein found in human milk, making up 20-25% of the whey proteins. The primary structure of this protein consists of many different amino acids representing a readily available source of essential amino acids as well as branched amino acids, which is important from the nutritional point of view [Kelleher et al. 2003]. Some studies indicate that α-lactalbumin has a relevant role in the absorption of ions: it is known that the human α-lactalbumin is complexed with Ca^{2+} ion and can also have Zn^{2+} ion as a binder. Although the maximum amount of calcium ions complexed with α-lactalbumin in breast milk is only 1% of the total calcium content of human milk [Ma et al. 2020; Lönnerdal and Glazier, 1985], it is possible that α-lactalbumin may have a positive effect on the absorption of other minerals, possibly by the formation of peptides that facilitate the absorption of divalent cations. Kelleher et al. 2003 found that infant formula supplemented with bovine α-lactalbumin increases the absorption of zinc and iron in young rhesus monkeys, but there are still no concrete studies that relate the effect of human α-lactalbumin with mineral absorption in breastfed infants. Besides the nutritional function already discussed, new studies have been targeted at the analysis of the antimicrobial potential of α-lactalbumin. Polypeptides obtained after exposure of α-lactalbumin to proteases commonly found in the gastrointestinal tract have antimicrobial activity against *Escherichia coli*, *Klebsiella pneumoniae*, *Staphylococcus aureus*, *Staphylococcus epidermidis*, *Streptococci* and *Candida albicans*. Since the primary structure of the human α-lactalbumin is similar to that of a monkey, it can be speculated that proteolysis of the human α-lactalbumin could produce the same antimicrobial peptides conferring protection to infants during the lactation period (Chang et al. 2019; Golinelli et al. 2014).

Table 4. Examples of proteins present in human milk and colostrum, and their functions in the human newborn. (Golinelli et al. 2014)

Classification	Examples	Function in newborn
Enzymes	Lysozyme	Bacterial
	Lipase	Hydrolysis of fats
	Sulfhydryl Oxidase	Oxidation of Sulfhydryl groups, Regulation of Enzyme activity
	Glutathion Peroxidase	Selenium alloy facilitating its release for infant
Binders	Lactoferrin	Bactericidal, Possible role in intestinal iron absorption
	Haptcorrina	Possible bacteriostatic, Possible role in the absorption of Vitamin B12
	Folate binding protein	Possible role in the uptake of this vitamin serum, Possible role in the intestinal absorption
Nutritional	Lactalbumina	Rich source of amino acids Lactose synthesis
Protection	Immunoglobulins	Act as antibodies, such as IgA, IgG, etc.
	Fibronectin	Facilitating the training of particles by phagocytic cells
	Lactoferrin	Bacteriostatic- competes with siderophilic bacteria by ferric ion

Immunoglobulins

It is believed that immunoglobulins represent about 10-15% of whey proteins. There are five classes of antibodies; IgA, IgD, IgE, IgG and IgM. These maternal antibodies are of particular importance because the secretory immune system of the newborn only becomes mature several months after birth [Golinelli et al. 2011]. However, the mother's immunity against some pathogens can be transferred to the infant in the form of IgA, allowing the immature immune system of the newborns to be driven by the mother's acquired immunity. IgA is the most abundant antibody in milk. The IgA concentrations are elevated in early lactation (1.2 g/L) and maintained between 0.5 and 1 g/L until the 2nd year of lactation. IgA is produced by mammary gland cells; it is derived from the B cells of the small intestine and respiratory tract and then is transferred to the infant's digestive tract. IgA, due to its special molecular structure [Brandtzaeg, 2013], is resistant to intestinal proteolysis, by being absorbed by the endothelial membrane, therefore, it enters the systemic circulation and protects the neonate. The excretion of intact IgA was noted in breastfed neonates, and the amount of these proteins in the feces decreased according to their concentration in breast milk over the period of lactation [Ramsay et al. 2005]. IgA antibodies against pathogenic bacteria such as *Escherichia coli*, *Vibrio cholerae*, *Haemophilus influenzae*, *Streptococcus pneumoniae*, *Clostridium difficile* and *Salmonella*, and antibodies against viruses such as rotavirus, cytomegalovirus, HIV and influenza virus, and in addition antibodies against yeast such as *Candida albicans* were also found in breast milk, showing the amplitude of the defense system (Luna et al. 2020; Golinelli et al. 2011).

Lactoferrin

Lactoferrin (LF) belongs to the family of iron-binding proteins and has antimicrobial and immunotrophic function, being found in higher concentrations within the serum of human milk (Chang et al. 2020). Due to this fact, a high proportion of the iron present in breast milk is bound to LF, which facilitates the mineral uptake by the intestinal cells. LF is relatively resistant to proteolytic degradation in the gastrointestinal tract when compared to other milk proteins, such as casein, etc. hence facilitating the absorption of LF from milk by the neonate. The peptides resulting from LF proteolysis also have antibacterial activities (Luna et al. 2020; Wakabayashi et al. 2006). LF is absorbed within the intestine through specific membrane receptors, localized in intestinal cells. When orally administered, LF stimulates the immune response both locally and systemically, playing an important role in absorption of nutrients and also stimulating the proliferation of endothelial cells in the intestine and the growth of lymphoid follicles associated with the intestine (Chang et al. 2020; Aly et al. 2013). This property suggests the possibility of using LF in premature infants and in patients with intestinal diseases. LF controls the acceptable composition of intestinal microflora, by suppressing the growth of pathogenic bacteria and promoting the proliferation of Lactobacillus and Bifidobacterium. The newborns fed with artificial diets develop a harmful intestinal microflora (Enterococcus, Enterobacter, Bacteroides, Escherichia). The non-pathogenic microflora ensures a low pH, produces some vitamins, increases the activity of Natural Killer Cells (NK), T lymphocytes and macrophages, promotes the production of protective immunoglobulins and decreases the risk of allergies [Chang et al. 2019; Ushida et al. 2006]. In studies performed on rats, LF showed a protective effect in cases of bacteremia and endotoxemia. This protein stimulated the activity of cells of the reticulo-endothelial system and promoted my elopoiesis, thereby eliminating bacteria. In a model of experimental endotoxemia, this protein inhibited the activity of pro-inflammatory cytokines, nitric oxide, and reactive forms of oxygen. LF may also promote differentiation of T and B cells from immature precursors and increase the activity of natural killer cells (NK) and lymphokine-activated killer cells (LAK) [Wu et al. 2011]. It also provides protection against the toxicity of reactive oxygen radicals, and this property may be particularly relevant when the infant feeding is based on modified cow milk, containing iron mineral, since it is a source of free radicals. Together, these experimental studies support the idea that natural human milk has the best nutritional value for the newborn (Chang et al. 2019; Kunz et al. 2000). Supplementation of artificial food for newborns with LF seems to strongly enhance the protection and immunity during this category of food, so much so that the commercially available infant feeding formulas for newborns in the United States and Japan are all supplemented with LF [Chang et al. 2020; Aly et al. 2018].

ENZYMES

Lysozyme

One of the major components of the human milk whey fraction is lysozyme, an enzyme capable of degrading by hydrolyzing the glycosidic bonds of type β (1→4) between the N-acetylmuramic acid (NAM) and N-acetylglucosamine (NAG), the peptidoglycan cell wall of Gram-positive bacteria. Recent studies have shown that the addition of recombinant human lysozyme to chicken feed would function as a natural antibiotic, possibly suggesting that it could replace currently used antibiotic drugs. Lysozyme has also been shown to kill gram-negative bacteria *in vitro*, in a synergetic action with lactoferrin (Chang et al. 2019; Lee-Huang et al. 2005). By binding to lipopolysaccharide and removing it from the outer cell membrane of bacteria, lactoferrin will allow lysozyme to access and degrade the inner proteoglycan matrix of the membrane, thereby killing the microorganism. Lysozyme has also been shown to inhibit the growth of HIV *in vitro*, but in case of human milk, it may act on the free virus and not on cell-associated virus. The mechanism of antiviral activity is not yet known (Golinelli et al. 2014; Goldman et al. 2020).

Lactoperoxidase

Lactoperoxidase, in the presence of hydrogen peroxide (formed in small quantities by cells), catalyzes the oxidation of thiocyanate (part of saliva), forming hypothiocyanate, which can kill both gram-positive and gram-negative bacteria. Thus, lactoperoxidase in human milk may contribute to the defense against infection already within the mouth and upper alimentary canal. Human milk contains active lactoperoxidase, but its physiologic significance is not yet known (Shin et al. 2001; Goldman et al. 2020).

Lipase

Lipase present in human milk is Bile salt dependent (LDSB), it has a wide spectrum of functions allowing efficient use of cholesterol esters, mono, diand triglycerides, fat-soluble vitamins, long-chain fatty acids (> C18) lipoamides present in milk, either soluble and micellar [Chen et al. 1994]. In newborns, particularly preterms, the role of the bile-salt stimulated lipase is very important and accounts for 30-40% (approximately) of the lipid digestion as they have low lipase enzyme activity and poor lipid utilization [Golinelli et al. 2014; Chang et al. 2019].

Cytokines and Hormone

Cytokines are small and soluble glycoproteins, which act in an autocrine or paracrine manner by binding cellular receptors, on cascade operation, leading to the development and functioning of the immune system of the newborn. Human milk contains a number of pro-inflammatory cytokines, such as interleukin (IL) 1β, IL-6, IL- 8, tumor necrosis factor α, β factor and transforming growth factors (both TGF β1 and TGF β2) and TNF α and anti-inflammatory cytokines, such as IL-10 (Ustundag et al. 2005). Although all of these cytokines are immunomodulatory, as already pointed out, it seems that the overall effect of these factors on milk is to lessen the anti-inflammatory response in neonates. Despite its benefits an exaggerated inflammatory response results in reducing absorption and damaging the infant's intestine [Field, 2005; Luna et al. 2021]. Although these cytokines are present in low concentration (picograms), however, their relative concentration is higher in colostrum, after being reduced on the 21st day. These physiological modifications of the cytokine profile in different periods of lactation seem to be related to the required changes in the immune system of infants and neonates' needs for these cytokines. Various hormones are also present in the human milk such as cortisol, somatostatin, insulin, thyroid hormones, lactogenic hormones, oxytocin, prolactin, ghrelin adiponectin and leptin. Human milk also contains substances that modulate growth, such as Epidermal Growth Factor (EGF), Nerve Growth Factor (NGF), Growth Factor Similar to Insulin (IGFs), and interleukins. The transforming growth factors (TGF-alpha and TGF-beta) and colony-stimulating factor granulocyte (G-CSF) were also detected in human milk (Golinelli et al. 2014). These growth factors are secreted by the epithelial mammary gland cells, activated macrophages by lymphocytes (mainly T cells), or by neutrophils in milk. Some peptides such as the growth factors; the Epidermal Growth Factor (EGF), Growth Factor Releasing Hormone (GHRF) and Insulin- Like Growth I (IGF-I) are present in milk, and when absorbed, can influence the metabolism and also promote the growth and differentiation of various organs and tissues of the neonate. It appears that these growth factors protect cells against toxic substances and toxins and reduce the chance of neonatal inflammatory disease. Fibronectin is a protein that is involved in phagocytosis, and is present in human milk and levels of this protein in serum are higher in breast-fed infants than those fed with commercial infant formulas. Comparison of fibronectin isolated from milk and the one present in the plasma showed that they were both very similar, and that fibronectin is ingested intact from colostrum (Field, 2005; Ma et al. 2020).

Bacterial Composition of Human Milk

Culture-dependent research has shown that human milk collected aseptically from healthy mothers harbors viable commensal bacteria at concentrations of roughly log 3 colony-forming units (CFU)/mL. However, depending on the culture methodology and/or geographical factors, variations in bacterial concentrations (i.e., log 1–5 CFU/mL) (Jost et al. 2013) and diversity have been observed. In contrast to the variations in components such as proteins, little is known about the impact of factors such as lactational or gestational age on the bacterial composition of human milk (Beasley and Saris, 2004). A decrease in bacterial concentration over lactation time has been reported; this could be due, in part, to dilution by greater milk production, causing the net intake of bacteria by the infant to remain the same. However, this decrease yet to be confirmed by other studies (Jimenez et al. 2008). The bacteria most frequently detected by culture and subsequent viable strain isolation are members of the facultative anaerobic genera *Staphylococcus* and *Streptococcus* spp., mainly *Staphylococcus epidermidis* and *Streptococcus salivarius*, followed by *Propionibacterium* and *Enterococcus* spp. (Khodayar-Pardo et al. 2014). Furthermore, many other subdominant microbes, including other members of the Actinobacteria, Bacteroidetes, Firmicutes and Proteobacteria phyla, have also been isolated from the milk collected from healthy mothers (Alp and Aslim, 2010; Godmman et al. 2020). Due to the potentially beneficial characteristics of *Lactobacillus* and *Bifidobacterium* spp., numerous studies have successfully isolated these species from human milk using selective and elective culture media. Although, more than 200 bacterial species have been isolated from human milk till date except *Bifidobacterium* spp., the only viable gut-associated obligate anaerobes detected were *Peptostreptococcus* and *Veillonella* spp. (Perez et al. 2007). Sharma et al. 2016 isolated 8 *Lactobacillus* strains, out of which two isolates; *Lactobacillus crustorum* F11 and *Lactobacillus paraplantarum* KM1 have shown highly beneficial probiotic characteristics along with their antagonistic activity.

However, culture-independent molecular methods (e.g., quantitative polymerase chain reaction, denaturing/temperature gradient gel electrophoresis, cloning and 16S rRNA gene sequencing, and pyrosequencing) have allowed the detection of total bacterial concentrations of approximately log 6 genome equivalents per milliliter in human milk, as well as DNA belonging to major gut-associated obligate anaerobes not detected previously by culture methods. Consequently, members of the Bacteroidetes phylum (i.e., the Bacteroides group and therefore the genera Bacteroides and Prevotella) and a number of other members of the Clostridia class have now been identified, including the genera *Blautia*, *Clostridium*, *Dorea*, *Eubacterium* and *Ruminococcus* (Collado et al. 2009). Also, major butyrate producers, which are important for colonic health, have been identified, including *Coprococcus*, *Faecalibacterium*, *Roseburia* and *Subdoligranulum* (Godman et al. 2020; Cabrera-Rubio et al. 2012). Both culture-dependent and culture-independent methods, however, are associated with advantages and limitations. One important question is whether the array of gut-associated obligate anaerobes detected by molecular culture-independent methods escaped culture due to their fastidious growth requirements, or perhaps they represent dead cells and/or parts thereof. The latter would suggest that human milk might not be a suitable medium for their growth due to specific microbe–microbe interactions, i.e., immunological and antimicrobial factors present in human milk. In addition to possible geographical factors, the differences in human milk bacterial diversity seen between studies largely result from different methodologies including aseptic sampling procedures, selection of appropriate culture media, and incubation conditions. In case of molecular methods such as pyrosequencing, variations occur due to DNA extraction and 16S rRNA gene sequencing procedures (e.g., primer choice, sequencing depth, and raw sequencing data processing). In most studies, only a few bacterial populations of special interest were targeted; primarily *Bifidobacterium* and *Lactobacillus* spp. However, there is great potential to apply or develop culturing methods to detect a wider range of microbial targets including gut-associated obligate anaerobes, which are highly fastidious.

These approaches could be complemented by advanced methods such as ethidium monoazide bromide polymerase chain reaction, which make it possible to distinguish between inactive and active bacteria (Jeurink et al. 2013; Chang et al. 2020).

Origin of Human Milk Bacteria

The detection of bacterial DNA belonging to an array of obligate gut-associated anaerobes that are unlikely to proliferate outside their host raises the question of their origin. Indeed, the path of entry of bacteria into the mammary gland and human milk remains debatable; and it is hypothesized that bacteria may enter the mammary gland either from within, via a bacterial enteromammary pathway or by contamination from outside or a combination of both (Figure 5). "Contamination from outside may occur by skin contact and transfer of the neonatal oral microbiota (possibly acquired from maternal skin and oral microbiota) during breastfeeding" (Ramsay et al. 2005). This might explain the predominance in human milk of Staphylococcus (e.g., *S. epidermidis*) and *Streptococcus* spp. (e.g., *S. salivarius*), which are typical commensals of the skin and oral microbiota, respectively. *Lactobacillus* spp. in human milk could be acquired orally by the neonate during vaginal delivery, as this genus predominates the vaginal microbiota (e.g., *L. crispatus*). However, the strains of *L. crispatus* that have been isolated occasionally from human milk were not identical to vaginal strains of *Lactobacillus* spp., as shown by strain isolation and random amplified polymorphic DNA analysis fingerprinting, as well as by denaturing gradient gel electrophoresis fingerprinting and clone library analyses (Perez et al. 2007). Similarly, gut-associated obligate anaerobes may be transferred to the neonatal oral microbiota and human milk after contact with the perineal/anal microbiota during vaginal delivery. However, 16S rRNA gene clone library generation showed distinct differences between gut and oral microbiota. These observations are further supported by the fact that the bacterial composition of human milk expressed before and after labor, likewise as from women who delivered vaginally and by caesarean section, has been reported to be similar (Martın et al. 2004) which strongly supports the hypothesis of bacterial enteromammary pathway. Recently, the presence of a bacterial enteromammary pathway has been hypothesized and accordingly, it is believed that bacteria may translocate the maternal gut by internalization in leukocytes (e.g., dendritic cells) and subsequently circulate to the lactating mammary gland via the lymphatic and blood circulation.

What is likely to be the strongest support for the bacterial enteromammary pathway can be found in a recent study that isolated the same strain of *Bifidobacterium breve* from maternal feces, human milk, and corresponding neonatal feces. Mother-to-infant transmission of bifidobacterial strains was also reported by Makino et al. 2011. Other studies have shown that probiotic *Lactobacillus* strains (*L. salivarius* CECT5713, *L. fermentum* CECT5716, *L. gasseri* CECT5714, and *L. reuteri* ATCC 55730) administered orally to the lactating mother as an alternative to antibiotic treatment against infectious mastitis could be recovered from human milk (Abrahamsson et al. 2009). Since viable anaerobes were isolated at subdominant levels from human milk, larger numbers of samples and isolates and the development of culture methods that allow a lower detection limit will be required in future research to estimate the importance of and gain further insight into mother–neonate transfer quantitatively. Moreover, the complementation of culture with 16S rRNA gene pyrosequencing has made it possible to identify a range of gut-associated obligate anaerobes that are potentially present in human milk. It is very possible that new generations of probiotics will be identified within this new reservoir of microbes (Takako et al. 2020; Ismail et al. 2012).

Impact of Human Milk Bacteria on Neonatal Gut Microbiota

With a culturable bacterial density within the range of log 1–5 CFU/mL and an estimated daily ingestion of roughly 800 mL, human milk represents a source of log 4–8 viable bacteria/day that are able to influence neonatal gut microbiota establishment (Martın et al. 2004). Thus, it is expected that bacteria present in human milk are able to survive digestion and to act as pioneer bacteria and that the composition of the neonatal gut microbiota reflects, to a particular extent, that of human milk. This is supported by recent studies on population levels of facultative anaerobes such as *Staphylococcus*, *Streptococcus*, *Lactobacillus*, *Propionibacterium*, *Rothia*, *Escherichia*, and *Enterococcus spp.* and more importantly, gut-associated obligate anaerobes such as *Bifidobacterium*, *Bacteroides*, *Parabacteroides*, *Blautia*, *Clostridium*, *Collinsella* and *Veillonella* spp. in both ecosystems using culture and/or molecular methods(Consales et al. 2020; Gronlund et al. 2007). "More specifically, using culture, isolation and molecular fingerprinting methods, identical strains belonging to the genera *Bifidobacterium*, *Lactobacillus*, *Enterococcus* and *Staphylococcus* have been detected in neonate faeces, maternal milk and eventually, maternal faeces, which are assumed to be vertically transferred from mother and neonate via breastfeeding" (Figure 6) (Arboleya et al. 2010). This mechanism could, in part, explain differences in gut microbiota establishment between breast and formula-fed neonates. Exclusively breastfed neonates have been shown to acquire a relatively simple and stable microbiota during the first to second week of life, predominated by *Bifidobacterium* spp. In contrast, the microbiota of formula-fed neonates has been reported to be more diverse but less stable, harbouring less *Bifidobacterium*, *Lactobacillus* and *Staphylococcus* spp. and more *Bacteroides*, *Enterococcus*, *Streptococcus* spp., as well as members of the Enterobacteriaceae family and Clostridia class. Despite such differences in early microbiota establishment, the cause and consequence relationships remain largely unknown and difficult to link to intestinal colonization by specific bacterial groups. Low diversity of the gut microbiota is believed to be disadvantageous and has been associated with several diseases (Ismail et al. 2012). This suggests the importance of synergism with other human-milk components that influence the survival and colonization of specific populations, as well as the timing of first contact, in directing the establishment of the microbiota and gut and immunity maturation (Golinelli et al. 2014).

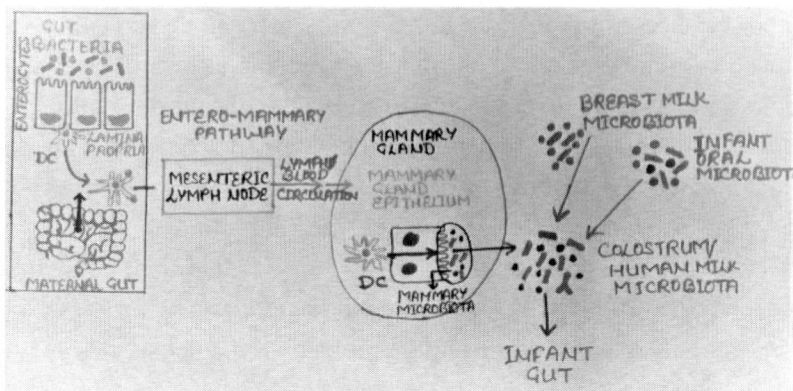

(Source: Fernández et al. 2012)

Figure 5. Potential sources of the bacteria present in human colostrum and milk. DC: dendritic cell.

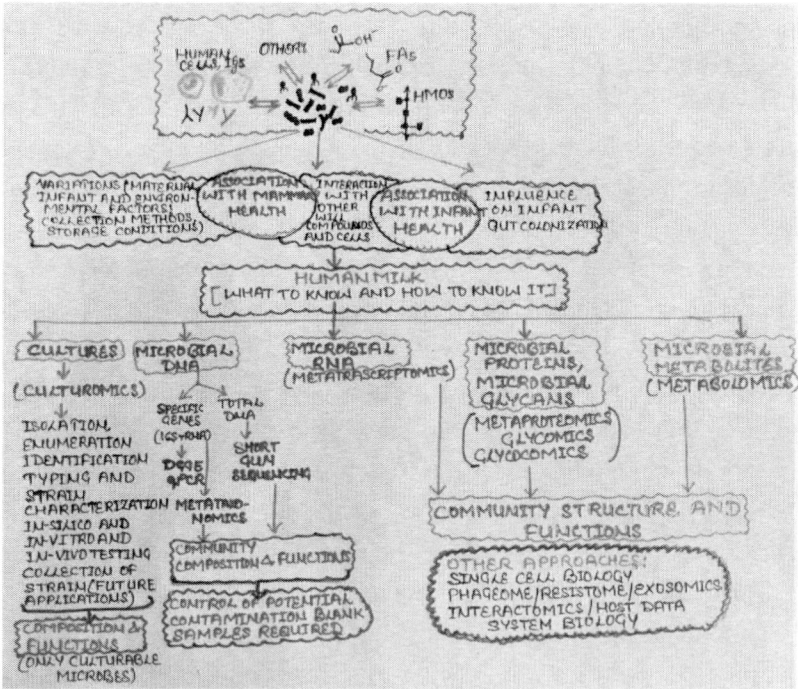

(Source: Ruiz et al. 2019)

Figure 6. Schematic overview of what we need to know about the human milk microbiome and the main tools that can be applied to achieve their study.

Bacteria of the Genus *Lactobacillus* spp.

There are more than 200 strains of bacteria in breast milk, of which the most important are *Lactobacillus* and *Bifidobacterium*. *Lactobacillus* bacteria (belonging to lactic acid bacteria) have the ability to break down lactose and other simple sugars into lactic acid (Morphology of *Lactobacillus* strain is given in Figure 8) [Jurkowski et al. 2012; Jost et al. 2013]. The microbiological profile of breast milk varies among individuals, both in quality and quantity. Some strains are constantly present in breast milk, while others such as *Lactobacillus* are characterized by variability and are not present in milk samples [Jost et al. 2013]. For example, Sinkiewicz and Ljunggren, 2008 isolated *L. reuteri* from the milk of approximately 15% of breastfeeding women. *L. salivarius*, *L. fermentum* and *L. gasseri* were dominant in this group. Tests showed a lower frequency of bacteria of the genus *Lactobacillus* in milk samples from women who underwent antibiotic therapy or cesarean delivery. A relationship was also found between the reduced detection of *L. salivarius* and the use of anesthesia during delivery [Soto et al. 2014]. The spectrum of *Lactobacillus* groups is narrow in breast milk and infant feces. The composition of bacterial strains is specific and includes only a small number of Lactobacilli strains [Golinelli et al. 2014; Martin et al. 2007]. Human milk is a crucial source of LAB in shaping the intestinal microflora of breast-fed infants. A study by Matsumiya et al. showed that less than one-fourth of newborns with vaginal delivery were acquired by vaginally derived lactic acid bacteria. After one month, these bacteria were replaced with lactic acid bacteria from breast milk [Matsumiva et al. 2002]. *Lactobacillus* composition is similar within a mother-child pair and contains small number of Lactobacilli strains [Heikkilä, and Saris, 2003]. Studies indicate that lactic acid bacteria in the gastrointestinal tract are found only in 74% of infants in the first months of life. Ahrné et al. isolated one strain of lactic acid bacteria from 37% of infants; 26% had two strains, and 11% had three or more strains. In this study, most babies were fed breast milk. Handa and Sharma, 2016 isolated *Lactobacillus crustorum* bacteria from human milk and studied its antimicrobial potential by production of bacteriocin (Grönlund et al. 2003). *Lactobacillus* bacteria were more often isolated from the stools of infants receiving breast-milk than from weaned infants [Ahrné et al. 2005]. Research by Martin et al. 2012 indicated the presence of the same specific *Lactobacillus* strains in breast milk and breastfed infant faeces. Compared to that in formula-fed infants, the intestinal microbiota of breastfed infants was richer in Lactobacilli and Bifidobacteria. At the same time, a reduced number of Veillonellaceae, Enterococcaceae, Streptococcaceae and Lachnospiraceae was observed [Ma et al. 2020]. Many studies confirm the health-promoting effects of *Lactobacillus* bacteria derived from breast milk on the baby's body. Studies have shown the effectiveness of disease prevention by *L. reuteri*, *L. rhamnosus* and many others [Urbanska et al. 2014]. The probiotic potential of *Lactobacillus* bacteria isolated from breast milk is similar to that of strains used in probiotic products used in medicine (Martin et al. 2005). Their safety is appreciated because of their natural origin, anti-infectious and immunomodulatory properties [Chang et al. 2020; Arroyo et al. 2010].

Potential Factors Influencing the Milk Microbiome

It is known that genetic factors, mode of delivery, maternal nutrition, time of day, lactation stage and geographical region influence human milk composition and considerable inter-individual variation has also been reported [Aly et al. 2020; Quinn et al. 2012]. Similarly, all the factors that could modulate the microbiota of the mother's skin, oral cavity, vagina and intestine, and the microbiota of the infant are potentially able to modulate the human milk microbiota. Therefore, the lactation period, maternal dietary habits and nutritional status, mode of delivery, gestational age, geographical location, and the use of antibiotics or other medicines can all have an influence on the milk microbiota. Higher microbial diversity has been reported in colostrum samples than in mature milk. Lactation stage has been described as an element influencing milk microbes. Initially, the microbiota is dominated by *Weissella*, *Leuconostoc*, *Staphylococcus*,

Streptococcus, and *Lactococcus* spp. [Ma et al. 2020; Cabrera-Rubio et al. 2012]. Later, the microbiota contains high levels of *Veillonella*, *Prevotella*, *Leptotrichia*, *Lactobacillus*, *Streptococcus* spp., and increasing levels of *Bifidobacterium* and *Enterococcus* spp [Khodayar-Pardo et al. 2014]. The mode of delivery affects human milk microbiota composition. High microbial diversity and high prevalence of *Bifidobacterium*and *Lactobacillus* spp. are found in colostrum and milkfollowing vaginal delivery, whereas the contrary is observed following cesarean delivery [Hoashi et al. 2015] although other studies did not report differences in microbial profiles based on gestation, mode of delivery or infant gender [Urbaniak et al. 2016]. Breast-milk microbiota composition is also influenced by gestational age, with significant differences between term- and preterm-delivered mothers. Lowercounts of *Enterococcus* spp. in colostrum and higher counts of*Bifidobacterium* spp. in milk have been detected in samples from mothers with term deliveries [Khodayar-Pardo et al. 2014]. In addition, changes in the milkmicrobiota composition are associated with maternal physiological status including obesity, celiac disease, and human immunodeficiency virus (HIV)-positive status [Collado et al. 2012]. Obesity is reflected within the levels of *Bifidobacterium* spp. and cytokines in human milk, as well as increased *Staphylococcus* spp., leptin, and proinflammatoryfatty acid levels and reduced microbial diversity [Panagos et al. 2016]. Mothers with celiac disease have reduced levels of cytokines, *Bacteroides* spp. and *Bifidobacterium* spp. in their milk [Olivares et al. 2015]. Finally, the milk from HIV-positive women from Africa has been found to point out higher bacterial diversity and improved prevalence of *Lactobacillus* spp. than the milk of non-HIV positive women[Gonzalez et al. 2013]. Analysis of the human milk microbiota shows that, in general, *Staphylococcus* and *Streptococcus* spp. as well as lactic acid bacteria strains are present in milk [Jost et al. 2014], but their relative amounts and the presence of other bacteria could be dependent on geographical location [Chang et al. 2019; Gonzalez et al. 2013]. However, large-scale studies with breast-milk samples from different geographic locations are needed. It is also evident that perinatal use of antibiotics has an effect on the maternal microbiota, including human milk microbiota, affectingthe prevalence of *Lactobacillus*, *Bifidobacterium*, and *Staphylococcus* spp. and decreasing the abundance of *Bifidobacterium*, *Staphylococcus*, and *Eubacterium* spp. in milk samples [Witt et al. 2014]. Chemotherapy has also been associated with alteration in the milk microbiome and with a reduction in bacterial diversity [Urbaniak et al. 2014]. Studies on the impact of maternal diet on milk microbiota are few in number, but it's likely that nutritional habits, which modulate intestinal microbiota and human milk nutritional composition, may exert changes in the milk microbiota. For example, shared microbial features have been reported between bacteria present in local foods and other fermented foods, as well as mother infant microbial gut and breast-milk [Albesharat et al. 2011]. Moreover, diet clearly influences the lipid profile of human milk, modulating the concentration of long-chain polyunsaturated fatty acids and the ratio of u-3 to u-6 [Nishimura et al. 2014]. These fatty acids have immunomodulatory properties in nursing children and it has recently been demonstrated in animal models that they are able to modulate gut microbiota composition in early life [Pusceddu et al. 2015]. The consumption of probiotics and prebiotics during pregnancy could influence the human milk microbiota, but more studies are required to document the potential transfer of gut bacteria to the mammary glands and the impact of specific strains of bacteria. On the other hand, it has been reported that non-nutritive sweeteners such as saccharin, sucralose, and acesulfame potassium were present in 65% of breast-milk samples analyzed [Luna et al. 2021; Sylvetsky et al. 2015]. Those data suggest that maternal diet could modulate the bioactive compounds and microbes present in breast milk. Studies are urgently needed to investigate interactions among nutrients in the maternal diet and the breast-milk microbiota and their health effects on infants (Gómez-Gallego et al. 2018).

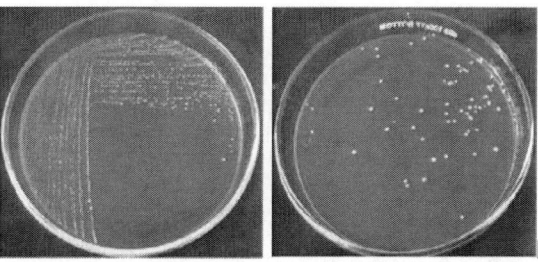

Figure 7. *Lactobacillus* strains isolated from human milk.

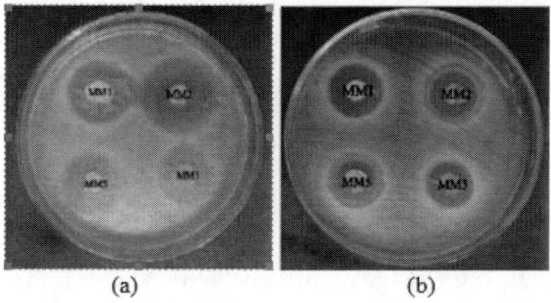

Figure 8. Bacteriocin producing *Lactobacillus* strains showing inhibition of growth of (a) *E. coli* and (b) *Staphylococcus aureus*.

Beneficial Properties of Human Milk Bacteria

Given the considerable bacterial diversity of human milk, interest in assessing the beneficial characteristics and thus, the probiotic potential of strains isolated from human milk has increased in recent years. The World Health Organization defines probiotics as "live micro-organisms which, when administered in adequate amounts, confer a health benefit to the host" (FAO/WHO, 2002). It has been shown that strains of *Bifidobacterium, Enterococcus, Lactobacillus, Lactococcus, Leuconostoc, Staphylococcus* and *Streptococcus* spp. isolated from human milk are able to inhibit the growth *in vitro* of gastrointestinal microbes that are potentially pathogenic (i.e., agar diffusion tests), including strains of *Bacillus, Clostridium, Cronobacter, Escherichia, Klebsiella, Listeria, Proteus, Pseudomonas, Salmonella, Shigella* and *Staphylococcus* spp.

Using culture-independent methods, the genus *Pseudomonas* has been detected at high relative amounts in human milk; whereas strains of this nonfastidious genus have rarely been isolated using culture dependent methods (West et al. 1979). The protective mechanisms of bacteria in human milk may, therefore, already be active, while the detected DNA may correspond to stressed, dead, or (partially) lysed cells by the action of antimicrobial components and/or weakness for competition against other bacteria of the human milk ecosystem. For instance, several strains of *Lactobacillus* sp., *Lactococcus lactis* as well as *Enterococcus faecalis* strain C901 isolated from human milk are capable of producing nisin and enterocin, respectively (Maldonado-Barragan et al. 2009). Similarly, protective mechanisms could also be active within the neonatal gut ecosystem and influence gut microbiota establishment, such as the competition for nutrients and intestinal mucosal binding sites (i.e., colonization resistance/competitive exclusion), immunomodulation (e.g., induction of secretory IgA, balancing pro and anti-inflammatory responses [e.g., interleukin-10]), enhancement of mucosal barrier function (e.g., reduction of permeability via exopolysaccharide production and induction of mucin production) and production of antimicrobial compounds (e.g., organic acids, hydrogen peroxide and bacteriocins) (Maldonado et al. 2012). Furthermore, considering the bacterial composition of human milk, *Veillonella* spp. and *Propionibacterium* spp. may play an important role in the establishment of a balanced trophic chain in the neonatal gut. These species metabolize lactate derived from lactose fermentation to propionate and acetate. These trophic functions may be transferred to the breastfed neonate and contribute to the establishment of balanced metabolic activities of the gut microbiota and may prevent disorders such as ulcerative colitis resulting from intestinal lactate accumulation (Chassard et al. 2014). Although human milk is a source of commensal bacteria, there have been case reports of diseases caused by opportunistic bacteria and pathogens such as *Brucella melitensis*, *Leptospira interrogans*, *Listeria monocytogenes*, *Salmonella enterica*, *Staphylococcus aureus* and *Streptococcus agalactiae* transferred from infected human milk (Olver et al. 2000).

Table 5. Antimicrobial peptides of Lactic acid bacteria

S. No.	Produce		Main Target Organism
1.	Enzymes I. Lactoperoxidase system II. Lysozyme (by recombinant DNA)		Pathogens and spoilage causing bacteria (milk and dairy product with hydrogen peroxide) Undesired Gram +ve bacteria
2.	Bacteriocins	I. Lantibiotics	Ribosomally produced peptides that undergo the extensive post-translational modification Small (<5 kDa) peptides containing lanthionine and methyl lanthionine Ia. Flexible molecules compared to Ib Ib. Globular peptides with no net charge or net negative charge
		II. Nonlantibiotics	Low molecular-weight (<10 kDa), heat stable peptides Formed exclusively by unmodified amino acids Ribosomally synthesized as inactive peptides that get activated by post-translational cleavage of the N-terminal leader peptide IIa. Anti-listerial single peptides that contain YGNGGVXC amino acid motif near their N termini IIb. Two peptide bacteriocins IIc. Bacteriocin produced by the cell's general sec-pathway
		III. Nonlantibiotics	High-molecular-weight (>30 kDa), heat labile proteins
		IV. Others	Complex bacteriocins carrying lipid or carbohydrate moieties, which appear to be required for activity Such bacteriocins are relatively hydrophobic and heat stable
3.	Low molecular weight metabolites (a) Revterin (b) Diacetyl (c) Fatty acid		Wide spectrum bacteria, yeast, mold Gram -ve bacteria Different bacteria
4.	Organic acid (a) Lactic acid (b) Acetic acid (c) Hydrogen peroxide		Putrefactive and Gram -ve bacteria, some fungi. Putrefactive bacteria, clostridia, some yeast and some fungi Pathogens and spoilage organisms especially in protein rich food

Antimicrobial Activity

Antagonistic properties are the ability of the probiotic strains to show antimicrobial activity against pathogenic bacteria by means of antimicrobial substance such as bacteriocin (Gupta et al. 2018) as well as other end products metabolites like organic acids and ethanol (Table 5) (Nallala et al. 2017). Inhibition of the growth of pathogen at refrigerated temperatures is possible by the production of hydrogen peroxide (Gupta et al. 2018). This attractive finding can be used in food production. Recently, Serrano-Niño et al. (2016), in their study found that all *Lactobacillus* spp. isolated from human milk showed positive inhibition to *S. aureus*, *E. coli*, *Salmonella* spp., and *Listeria* spp. Sharma et al. 2016 had explored the antimicrobial properties of *Lactobacillus* strains isolated from human milk and colostrum and observed high antimicrobial activity against *E.coli*, *S. aureus*, *C. perferingens* and *L. monocytogenes*. However, in a study which set out to determine potential probiotics isolated from human milk and colostrum, the antagonistic activity of the selected three Lactobacillus isolates L2, L4, L5 of human milk were found to inhibit the growth of enteropathogens. All the three isolates inhibited the growth of *E. coli*. The L2 isolate reported the strongest inhibitory activity against the gastro intestinal enteropathogens like *E. coli*, *Enterococcus faecalis*, *Pseudomonas fluorescence*, *Pseudomonas aerugenosa*, *Staphylococcus aureus*, *Salmonella typhimurium* and *Proteus mirabilis* while less antagonistic activity was observed against *Bacillus megaterium1684* and *Xanthomonas campestris*. The *Lactobacillus* spp. L2 exhibiting probiotic properties and strong inhibitory activity was selected for the optimization of bacteriocin production using various sources. Furthermore, the tested isolates (strains) were ready to inhibit the expansion of human enteropathogens (Gupta et al. 2018; Meera et al. 2012). A multitude of proteins in human milk have inhibitory activities against pathogenic bacteria, viruses, and fungi. Some of these proteins are likely to act independently, whereas others may act synergistically. There appears to be considerable redundancy with several components working on an equivalent pathogen; this means a multilayered defence system which can explain the lower prevalence of infection in breastfed infants than in formula-fed infants. Several antimicrobial activities have been ascribed to lactoferrin. Originally, it was believed that lactoferrin, being largely unsaturated with iron, could withhold iron from iron-requiring pathogens because of its high affinity for iron, thereby exerting bacteriostatic activity. Although this is often possible, several studies have also shown a robust bactericidal activity of lactoferrin against several pathogens, which isn't hooked in to the degree of iron saturation of lactoferrin. Some, if not all, of this activity may be the result of the formation of lactoferricin, a potent bactericidal peptide formed during the digestion of lactoferrin. Recent studies also showed that lactoferricin inhibits the attachment of enteropathogenic *E. coli* (EPEC) to intestinal cells (Chang et al. 2020; Izadpanah and Gallo, 2005), which appears to be mediated by the serine protease activity of lactoferrin. By degrading the protein structures of EPEC that are needed for the attachment and invasion of the bacteria, infection could also be blocked. It, therefore, appears that several activities of lactoferrin contribute to defence against bacterial infection. Lactoferrin was also shown *in vitro* to have activity against viruses such as HIV and fungi, such as *C. albicans* but the mechanisms behind these activities are not known.

BACTERIOCINS

Bacteriocins are ribosomally synthesized peptides with antibacterial activity produced by bacteria viz. *Lactobacillus, Enterococcus, Pediococcus, Streptococcus, Weisella, Carnobacterium*, etc. as given in Table 6. Bacteriocins can be post-translationally modified (PTM) or non-modified and are grouped into different classes (Alvarez Sieiro et al. 2012). For example, lantibiotics are modified bacteriocins comprising membrane-active peptides with thioether-containing amino acids lanthionine and β-methyllanthionine and are grouped in class I,. Most antimicrobial peptides are positively charged and smaller than 10 kDa (with the exception of the class III bacteriocins). Their small size, charge, and variation in hydrophobic and hydrophilic properties allow them to adhere to microbial cells and penetrate phospholipid membranes (Izadpanah and Gallo, 2005). Bacteriocins adhere to organisms via "docking molecules." Lipid II, for instance, is the docking molecule for several lantibiotics (mostly nisin-like). Mannose phosphotransferase proteins IIC/D is a docking molecule for sophistication IIa bacteriocins. The mode of actionis either pore formation, thus destabilizing the proton motive force, or inhibition of DNA, RNA and protein-synthesis (Cotter et al. 2013). Some bacteriocins exhibit a much broader spectrum of activity and may extend beyond the borders of bacteria to include protozoa, yeast, fungi, viruses, and eukaryotic cells (Figure 8) (e.g., cancer cells and spermatozoa; Chikindas et al. 2018). In their natural environment, bacteria produce bacteriocins to compete against other bacteria for nutrients. Complex environments with changes in growth conditions, nutrients, pH, water activity (Aw) and temperatures may have an impact on bacterial cell numbers, their metabolic activity, and bacteriocin production. In the gastro-intestinal tract (GIT), changes in food particles and fluctuation in spices, additives, salts, bile, digestive enzymes, etc. may have a negative impact on cell growth and bacteriocin production. Once secreted by the producer cells, the activity of bacteriocins could also sufferfrom their ability to stick to food particles and diffuse through the digesta, but also their stability at different pH values, resistance to digestive enzymes and proteolytic enzymes, and chemical interactions with particles and microbial cells within the GIT. It is, thus, extremely difficult to visualize the functioning of bacteriocins in the GIT. The production of nisin A, for instance, is regulated by a protein pheromone, via a two-component signal transduction system almost like quorum-sensing systems. In a fermenter, the addition of nisin to a culture stimulates the expression of nisin. This may very well be the situation in the GIT. An increase in bacteriocin levels, e.g., when adhered to lipids in the mucus layer, may have an effect on the expression of genes encoding bacteriocin production (Dicks et al. 2018). Bacteriocins are membrane active cationic peptides, and should thus even have an impact on mammalian cell membranes. Cinnamycin and duramycin bind phosphatidylethanolamine (PE), a substrate for phospholipase A2 and involved in inflammatory responses (e.g., vascular inflammation). The sequestering of PE by cinnamycin and duramycin may thus end in immune modulation through the indirect inactivation of phospholipase A2. Furthermore, by binding PE, the peptides could also be deposited in cellular membranes, thereby potentially changing biophysical membrane properties.

Table 6. Bacteriocin producing isolates and their sources

Source	Name of isolate	References
Sour dough	*L. reuteri*, *L. bavaricus*, *L. curvatus* and *L. Plantarum*	(Ganzale et al., 2004);
Rippened soft cheese	*C. piscicola*	(Herbin et al., 1997)
Cheese	*B. linenes* and *L. paracasei* subsp. *paracasei* BGUB9	Tolinaki et al., 2010)
Vacuum packed meat	*L. sake*	(Motta and Brandelli, 2003)
Commercial salad	*L. casei*, *L. plantarum*, *Pediococcus* sp. and *Lactobacillus* sp.	(Vescovo et al., 1996)
Sausages	*L. brevis* SB27 and *L. curvatus* SB	(Benoit et al., 1997)
Raw barley	*Lactobacillus* sp.	(Hartnett et al., 2002)
Chiken meat	*L. mesenteroides* and *L. plantarum*	(Martinez and Martinez, 2006)
Idli batter	*L. lactis* MTCC 3041	(Garcha and Singla, 2011)
Whey	*L. helviticus* G51 and *B. mycoides*	(Sharma and Gautam, 2008)
Barley beer	*L. paracasei*, *L. pentosus*, *L. plantarum* and *Lactococccus lactis* subsp. *lactis*	(Todorov et al., 2004)
Human hand	*Bacillus* sp. 8a	(Bizani and Brandelli, 2002)
Soil	*radiobacter* NA6	(Jabeen et al., 2009)
Tempeh	*Lactobacillus* sp.	(Kormin, 1998)
Yogurt	*L. plantarum*	(Shouny et al., 2012)
Som-fak	*Lactobacillus plantarum* PMU 33	(Noonpakdee, 2009)
Khameeara	*B. lentus*	(Sharma et al., 2006)
Burkina faso fermented milk	*Lactobacillus fermentum*, *Pediococcus* subsp. and *Leuconostoc mesenteroides* subsp. *Meseteroides*	(Savadogo et al., 2004)
Kimchi	*L. brevis* 925A	(Wada et al., 2009)
Marcha	*Bacillus* sp. AGI	(Sharma et al., 2011)
Gajani sikhe	*Lactobacillus* sp. SPY21	(Chin et al., 2001)
Boza	*Lactobacillus plantarum* ST194BZ,	(Todorov and Dicks, 2005)
Green olive	*L. fermentum* NM 332	(Mojgani et al., 2009)
Agro industrial wastes	*Bacillus megaterium* 19	(Khalil, 2009)
Chilli pickle	*Pediococcus acidilactici*	(Kaur and Balgir, 2004)
Goat meat	*E. faecium*	(Dutta et al., 1999)

These changes can cause altered ion channel functioning. In case of duramycin, this characteristic is exploited for the potential treatment of cystic fibrosis. Minimal cytotoxic effects of bacteriocins against human cell lines have been reported (Dreyer, 2018). Given the concentration of bacteriocin required to induce significant cytotoxicity, these levels would most likely not be present as a result of bacteriocins crossing the gut blood barrier (GBB). This, however, does not discount the possibility of bacteriocins accumulating in organs such as the liver and causing membrane damage.

(a)   (b)

(Source: Sharma et al. 2020)

Figure 9. (a) EPS production by *Lactobacillus paraplantarum* KM1, (b) The lyophilized polysaccharide fractions.

EXOPOLYSACCHARIDES PRODUCTION

Exopolysaccharides are long-chain polysaccharides produced extracellularly mainly by bacteria and microalgae. They consist of branched, repeating units of sugars or sugar derivatives with some non-carbohydrate substitutes such as acetate, pyruvate, succinate and phosphate (Figure 10). These sugar units are mainly glucose, galactose, mannose, N-acetylglucosamine, N-acetyl galactosamine and rhamnose, in variable ratios (Table 7). Bacterial polysaccharides synthesized and secreted into the external environment or synthesized extracellularly by cell wall-anchored enzymes could also be referred to as exopolysaccharides (Figure 10).

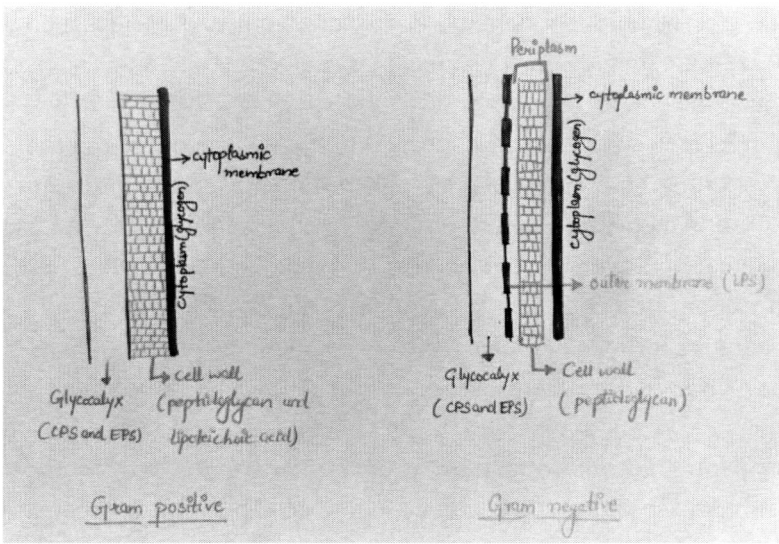

(Source: Looijesteijn et al. 2001).

Figure 10. Cell location of polysaccharides produced by gram-positive and gram-negative bacteria. CPS = Capsular polysaccharides (capsule), EPS = exopolysaccharides (slime layer).

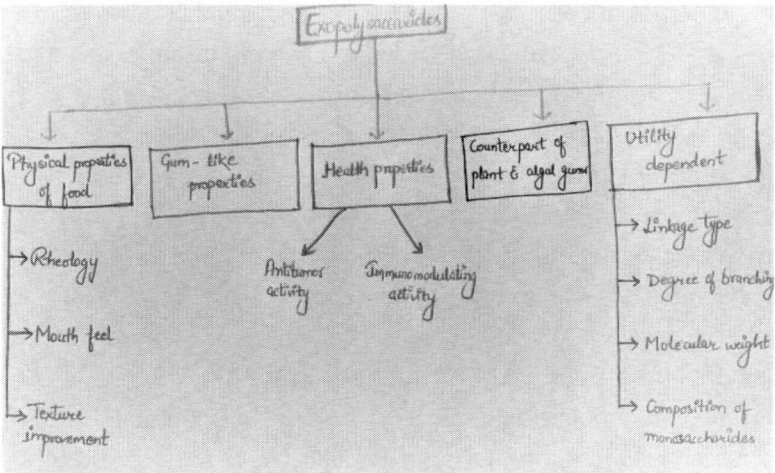

Figure 11. Wide spectrum activities of exopolysaccharides from lactic acid bacteria (Parvez et al. 2006).

The overwhelming diversity of bacterial polysaccharides allows for categorization based on chemical structure, functionality, relative molecular mass and linkage bonds (Nwodo et al. 2012; Sanchez et al., 2006). EPS aren't permanently attached to the surface of the microbial cell and are secreted into their surroundings during growth as loose slime. This distinguishes them from the structurally similar capsular polysaccharides, which remain permanently attached to the microbial cell surface. EPS play a vital role in protection of the microbes from adverse conditions such as dessication, nutrient shortage, toxic compounds, bacteriophages and osmotic stress (Looijesteijn et al. 2001). EPS play key role in initial adhesion and firm anchorage of the bacteria to solid surfaces, cation sequestration, biofilm formation, cellular recognition and pathogencity. Generally, the EPS aren't utilized as food by the bacteria producing it, but Streptococcus mutan and Streptococcus sobrinus are capable of degrading self-synthesized EPS where Streptococcus sobrinus degrades dextran and Streptococcus mutans utilizes the mutan produced by it (Sanchez et al., 2006; Patel et al. 2012). Exopolysaccharides are water soluble gums with novel and unique physical properties. EPS produced by food-grade microorganisms with GRAS (generally recognized as safe) status have gained a lot of importance as they are excellent alternatives to the polysaccharides of plant and animal origin.

Exopolysaccharides from carboxylic acid bacteria confer beneficial rheological and functional properties to foods such as yoghurts as natural thickening agents, giving the product appropriate viscosity and reducing syneresis. Due to their far better mouth-feel and longer retention time within the mouth, the exopolysaccharides are believed to boost flavour perception in viscous foods. The consistency and viscosity of the ultimate products are often improved with the assistance of EPS. Some EPS are shown to possess health-beneficial properties viz. immune stimulation, antiulcer activities, antitumoral activites and cholesterol-lowering activity (Figure 11) (Parvez et al. 2006). The exopolysaccharides produced by probiotic microbes enhance the improvement within the properties of the ultimate product due to the presence of probiotic organism as well as EPS secretions.

POTENTIAL STRATEGIES TO MODULATE THE MATERNAL BREAST-MILK MICROBIOTA

As mentioned above, diet is a powerful tool to alter gut microbiota. Therefore, dietary strategies could be devised to modulate the microbial composition in order to affect human physiology and reduce the risk of diseases related to imbalances of microbiota composition. A recent study [Ma et al. 2014], with Japanese macaques (Macaca fuscata) as animal model, has reported the impact of maternal diet on neonatal microbiome composition. This study showed how the maternal high fat diet affected infant microbiome composition and activity and also, the metabolic health. Thus, deciphering the contribution of specific gut bacteria and promoting nutrition and lifestyle counselling may open new tools to reduce the risk of diseases related to microbial composition shifts. Accumulating evidence shows that early dietary interventions and nutrition counselling would support health programming effects (immunological, metabolic and microbial effect) on adult health.

It has been demonstrated that specific probiotic strains are effective in the prevention and treatment of infectious diseases during early life [Taipale et al. 2016] and that they may reduce the risk of eczema in infantsin at-risk populations [Rautava et al. 2012]. In addition, a specific probiotic bacterium, *Lactobacillus reuteri*, has been detected in breast milk and infant feces when the mother consumed this probiotic [Abrahamsson et al. 2009; Chang et al. 2019]. In aplacebo-controlled study, probiotics ingestion during pregnancyand breastfeeding period has been described to modulate infant Bifidobacteria colonization and also to modulate breast milkmicrobiome [Gueimonde et al. 2006; Spartz 2021]. it has been recently, suggested that perinatal probiotic supplementation affects breast-milk composition in terms of microbes, including *Bifidobacterium* and *Lactobacillus* spp., and also affects other bioactive compounds like human milk oligosaccharides and lactoferrin [Mastroimarino et al. 2015]. Moreover, another similar study reported the beneficial effects of probiotic supplementation in vaginal birth, while no significant differences were found in milk samples from C-section deliveries, suggesting probiotic-specific dependent modulation depending on mode of delivery. A recent workshop report on the use of specific probiotics during the first 1500 days of life supports healthy perinatal life with lowered risk of infections and autoimmune problems later in life (Chang et al. 2019; Reid et al. 2016).

Table 7. Lactic acid bacteria and EPS monomers

Oraganism	EPS monomers	References
Lactobacillus acidophilus	Glucose & galactose	(Savagodo et al., 2004)
Lactobacillus acidophilus	Glucose	(Savagodo et al., 2004)
Lactobacillus delbrueckii ssp. *bulgaricus* NCFB 2483	Galactose, glucose, rhamnose & mannose	(Vincent et al., 2001)
Streptococcus macidonicus SCB6	D-glucose, galactose & N-acetylglusamine	(Vincent et al., 2001)
Lactobacillus plantarum	Glucose and mannose	(Tsuda and Miganoto, 2010)
Lactobacillus pentosus	Glucose and mannose	(Sanchez et al., 2006)
Lactobacillus plantarum MTCC 9510	Glucose and mannose	(Ismail and Nampoothiri, 2010)
Lactobacillus pentosus	Glucose and rhamnose	(Sanchez et al., 2006)

(Source: Wang et al. 2014)

NEW ADVANCES IN THE USE OF PROTEINS IN INFANT FORMULA

The presence of protective factors against infections, the absence of allergenic factors and the narrow affective mother child relationship are examples of the benefits of feeding the newborn with human milk, which are absent when the milk is replaced by infant formulas. The first important aspect observed when comparing human milk to infant formula is the qualitative and quantitative differences in certain nutritional components. Because of these differences, minimal recommendations are made for key compounds in human milk that are important for the development of the newborn [Aly et al. 2018; Gollinelli et al. 2014]. To compensate for the lower digestibility of the treated proteins found in infant formulas in comparison to proteins naturally found in human milk, the formula pattern must have a minimum protein content of 1.8 g/100 kcal (12 g/L) on a formula of 670 kcal/L. Thus, "the proteins made available are

25% higher than the average provided in the first 6 months of lactation. Formulae with higher protein content do not represent an advantage, as they overwhelm the metabolic and excretory functions of the neonate" [Dupont, 2003; Spatz, 2021]. Another important point is how the milk proteins are complexed with each other and divided into different fractions.

These characteristics are essential for their absorption, and other nutrients in the gastrointestinal tract, such as the minerals, zinc, iron and copper [Donangelo et al. 1991; Cimolai et al. 2020]. Cases of deficiency of these minerals are reported more frequently in infants fed formulas prepared from cow milk than in those fed with human milk, although the amount of these minerals in the formulas is generally higher [Lönnerdal, 1985; Luna et al. 2021]. From these results, the authors concluded that these minerals in human milk showed a higher efficiency of absorption (bioavailability), which seems to be due to the difference in distribution of these minerals in the fractions of milk (bovine and human) and infant formula. Moreover, differences in concentration of serum proteins such as lactoferrin and α-lactalbumin should also be considered, which has already been noted, also have an important role in the absorption and consequently the bioavailability of some of these minerals [Lönnerdal, 2004]. Conte-Junior et al. 2006 also described the presence of other proteins whose function appears to be related to absorption of minerals, but these proteins have not been fully identified and studied. The interest in producing recombinant proteins of human milk in addition to infant formulas has been growing in recent years [Aly et al. 2018; Lönnerdal, 2006]. Microorganisms and transgenic animals can be used for the production of proteins with biological activity. However, benefits from the addition of each protein in cells should be evaluated in animal models and ultimately in neonates [Chang et al. 2019; Hanson et al. 2001]. Another important point is to be careful with appropriate processing conditions so that the proteins added to formulas retain their biological activities. It is essential to use some processing conditions such as aseptic processing, sterile filtration, etc. to maintain the unchanged tertiary structures of proteins, and consequently preserve the biological activities. The importance of post-translational modifications should also be considered, because some proteins may require glycosylation and/or phosphorylation to present appropriate physiological activity [Bethell and Huang, 2004; Consales et al. 2020]. To summarise, knowledge of the composition of human milk and the factors that influence it has been increased considerably over the past two decades. Human milk is the best nutrition for the newborn because breastmilk is a complex fluid, rich in nutrients and in non-nutritional bioactive components (Gollinelli et al. 2014; Consales et al. 2020).

REFERENCES

[1] Abrahamsson, T. R., Sinkiewicz, G. & Jakobsson, T. (2009). "Probiotic lactobacilli in breast milk and infant stool in relation to oral intake during the first year of life." *Journal of Pediatr Gastroenterol Nutriton*, *49*, 349–54.

[2] Abrahamsson, T. R., Sinkiewicz, G., Jakobsson, T., Fredrikson, M. & Bj€orksten, B. (2009). "Probiotic lactobacilli in breast milk and infant stool in relation to oral intake during the first year of life." *Journal of Pediatr Gastroenterol Nutriton*, *49*, 349-54.

[3] Adkins, Y. & Lönnerdal, B. (2002). "Mechanisms of vitamin B12 absorption in breast fed infants." *Journal of Pediatr Gastroenterol Nutrition*, *35*, 192–98.

[4] Ahrné, S., Lönnermark, E. Wold, A. E., Aberg, N., Hesselmar, B., Saalman, R., Strannegard, I. L., Molin, G. & Adlerberth, I. "Lactobacilli in the intestinal microbiota of Swedish infants." *Microbes Infection*, *7*, 1256–1262.

[5] Albesharat, R., Ehrmann, M. A., Korakli, M., Yazaji, S. & Vogel, R. F. (2011). "Phenotypic and genotypic analyses of lactic acid bacteria in local fermented food, breast milk and faeces of mothers and their babies." *Syst Appl Microbiol*, *34*, 148-55.

[6] Alp, G. & Aslim, B. (2010). "Relationship between the resistance to bile salts and low pH with exopolysaccharide (EPS) production of Bifidobacterium spp. isolated from infants feces and breast milk." *Anaerobe*, *16*, 101–05.

[7] Alvarez, S., Patricia, Montalban-Lopez, M., Mu, D. & Kuipers, O. (2016). "Bacteriocins of lactic acid bacteria: extending the family." *Applied Microbiology and Biotechnology.*, 100-10.

[8] Aly, E., Darwish, A., López-Nicolás, R., Frontela, C. & Ros, G. (2018). *"Bioactive Components of Human Milk: Similarities and Differences between Human Milk and Infant Formula".*, 10.5772/intechopen.73074.

[9] Aly, E., Ros, G. & Frontela, C. (2013). "Structure and functions of lactoferrin as ingredient in infant formulas." *Journal of Food Research*, *2*, 25-36.

[10] Arboleya, S., Ruas-Madiedo, P. & Margolles, A. (2010). "Characterization and *in vitro* properties of potentially probiotic Bifidobacterium strains isolated from breastmilk." *International Journal of Food Microbiology*, *149*, 28–36.

[11] Arroyo, R., Martín, V., Maldonado, A., Jiménez, E., Fernández, L. & Rodríguez, J. M. (2010). "Treatment of infectious mastitis during lactation: Antibiotics versus oral administration of lactobacilli isolated from breast milk." *Clinical Infection Diseases*, *50*, 1551–58.

[12] Bagwe-Parab, S., Yadav, P., Kaur, G., Tuli, H. S. & Buttar, H. S. (2020). "Therapeutic Applications of Human and Bovine Colostrum in the Treatment of Gastrointestinal Diseases and Distinctive Cancer Types: The Current Evidence." *Front Pharmacology.*, 11, 01100. Published 2020 S"ep 11. doi:10.3389/fphar.2020.01100.

[13] Beasley, S. S. & Saris, P. E. (2004). "Nisin-producing *Lactococcus lactis* strains isolated from human milk." *Applied Environmental Microbiology*, *70*, 5051–53.

[14] Benoit, V., Lebrihi, A., Bernard, M. J. & Lefebvre, G. (1997). "Purification and partial amino acid sequence of brevicin 27, a bacteriocin produced by *Lactobacillus brevis* SB 27." *Current Microbiology*, *34*, 173-79.

[15] Bethell, D. R. & Huang, J. (2002). "Recombinant human lactoferrin treatment for global health issues: iron deficiency and acute diarrhea." *Biometals*, *17*, 337-42.

[16] Bizani, D. & Brandelli, A. (2002). "Characterization of a bacteriocin produced by a newly isolated Bacillus sp. strain 8 A." *Journal of Applied Microbiology*, *93*, 512-19.

[17] Bode, L. (2009). "Human milk oligosaccharides: prebiotics and beyond." *Nutrition Review.*, *67*, 183–91.

[18] Bourtourault, M., Buléon, R., Sampérez, S. & Jouan, P. (1991). "Effect of proteins from bovine milk serum on the multiplication of human cancerous cells." *C R Seances Soc Biol Fil.*, *185*, 319-23.

[19] Brandtzaeg, P. (2013). "Secretory IgA: Designed for anti-microbial defense." *Frontier Immunology*, *4*, 222.

[20] Cabrera-Rubio, R., Collado, M. C. & Laitinen, K. (2012). "The human milk microbiome changes over lactation and is shaped by maternal weight and mode of delivery." *American Journal of Clinical Nutrtion*, *96*, 544–51.

[21] Chang, R., Ng, T. B. & Sun, W. Z. (2020). "Lactoferrin as potential preventive and adjunct treatment for COVID-19." *International Journal of Antimicrobial Agents*, *56*, 106118. doi: 10.1016/j.ijantimicag.2020.1061.

[22] Chassard, C., de Wouters, T. & Lacroix, C. (2014). "Probiotics tailored to the infant: a window of opportunity." *Current Opinion Biotechnology*, *26*, 141–47.

[23] Chen, Q., Bläckberg, L., Nilsson, A., Stenby, B. & Hernell, O. (1994). "Digestion of triacylglycerols containing long-chain polyenoic fatty-acids *in vitro* by colipase- dependent lipase, and bile-salt stimulated lipase." *Biochim. Biophys. Acta*, *1210*, 239-43.

[24] Chikindas, M. L., Weeks, R., Drider, D., Chistyakov, V. A. & Dicks, L. M. (2018). "Functions and emerging applications of bacteriocins." *Current Opinion Biotechnology*, *49*, 23-28.

[25] Chin, H. S., Shim, J. S., Kim, J. M., Yang, R. & Yaan, S. S. (2001). "Detection and antibacterial activity of a bacteriocin produced by Lactobacillus plantarum." *Food Science and Biotechnology*, *10*, 335-42.

[26] Cimolai, N. (2020). "Applying Immune Instincts and Maternal Intelligence from Comparative Microbiology to COVID-19." *SN Compr. Clin. Med.*, *2*, 2670–83.

[27] Collado, M. C., Delgado, S. & Maldonado, A. (2009). "Assessment of the bacterial diversity of breast milk of healthy women by quantitative real-time PCR." *Letters in Applied Microbiology*, *48*, 523–28.

[28] Collado, M. C., Laitinen, K., Salminen, S. & Isolauri, E. (2012). Maternal weight and excessive weight gain during pregnancy modify the immunomodulatory potential of breast milk." *Pediatr Res*, *72*, 77-85.

[29] Consales, A., Crippa, B. L., Cerasani, J., Morniroli, D., Damonte, M. & Bettinelli, M. E. (2020). "Overcoming rooming-in barriers: a survey on mothers' perspectives." *Front Pediatr*, *8*, 53.

[30] Conte-Junior, A. C., Golinelli, L. P., Paschoalin, V. M. P. & Silva, J. T. (2006). "Development of protein fractionation technique in the serum of colostrum for bidimensional electrophoresis for identification by mass spectrometry (MALDI-TOF)." *Food*, *373*, 120-21.

[31] Cotter, P. D., Ross, R. P. & Hill, C. (2013). "Bacteriocins - a viable alternative to antibiotics." *Nat Rev Microbiol*, *11*, 95-105.

[32] Crosse, V. M. (1966). *The Premature Baby*, p. 143. Churchill, London.

[33] Damaceno, Q. S., Souza, J. P., Nicoli, J, R., Paula, R. L., Assis, G. B., Figueiredo, H. C., Azevedo, V. & Martins, F. S. (2017). "Evaluation of Potential Probiotics Isolated from Human Milk and Colostrum." *Probiotics and Antimicrobial Proteins*, *9*, 371–79.

[34] Dicks, L. M. T., Dreyer, L., Smith, C. & Van Staden, A. D. (2018). "Corrigendum: A Review: The Fate of Bacteriocins in the Human Gastro-Intestinal Tract: Do They Cross the Gut–Blood Barrier." *Frontiers in Microbiology*, *9*, 2297.

[35] Donangelo, C. M., Trugo, N. M., Mesquita, V. L., Rosa, G. & Da-Silva, V. L. (1991). "Lactoferrin levels and unsaturated iron-binding

capacity in colostrum of Brazilian women of two socioeconomic levels." *Brazil Journal of Medical Biology Research, 24*, 889-893.

[36] Dreyer, L. (2018). *The Ability of Antimicrobial Peptides to Migrate Across the Gastrointestinal Epithelial and Vascular Endothelial Barriers*. MSc thesis, Stellenbosch University, Stellenbosch.

[37] Drozdowski, L. A., Clandinin, T. & Thomson, A. B. (2010). "Ontogeny, growth and development of the small intestine: understanding pediatric gastroenterology." *World Journal of Gastroenterology, 16*, 787–99.

[38] Dupont, C. (2003). "Protein requirements during the first year of life." *American Journal of Clinical Nutrition, 77*, 1544-49.

[39] Dutta, S., Chakraberty, P. & Ray, B. (1999). "Production, purification and characterization of Enterocin Cal1, a bacteriocin of Enterococcus faecium Cal1, isolated from goat meat." *Indian Journal of Microbiology, 39*, 15-22.

[40] FAO/WHO. *Report of a Joint FAO/WHO Working Group on Drafting Guidelines for the Evaluation of Probiotics in Food*. London, Ontario, Canada; 2002.

[41] Field, C. J. (2005). "The immunological components of human milk and their effect on immune development in infants." *Journal of Nutrition, 135*, 1-4.

[42] Ganzle, M. G. (2004). "Reutericyclin: biological activity, mode of action and potential application." *Applied Microbiology Biotechnology, 64*, 326-32.

[43] Garcha, S. & Singla, A. (2011). "Biopreservative potential of Lactococcus lactis MTCC 3041." *International Journal of Food and Fermentation Technology, 1*, 93-98.

[44] Gazzolo, D., Bruschettini, M., Lituania, M., Serra, G., Santini, P. & Michetti, F. (2004). "Levels of S100B protein are higher in mature human milk than in colostrum and milk-formulae milks." *Clinical Nutrition, 23*, 23-26.

[45] Godhia, M. & Patel, N. (2013). "Colostrum - Its Composition, Benefits As A Nutraceutical: A Review." *Current Research in Nutrition and Food Science Journal, 1*, 37-47.

[46] Goldman, A. S., Chheda, S., Keeney, S. E. & Schmalstieg, F. C. (2020). "Immunology of Human Milk and Host Immunity". *Fetal and Neonatal Physiology.*, *2011*, 1690–701.

[47] Golinelli, L. P., Conte-Junior, C. A., Paschoalin, V. M. F. & Silva, J. T. (2011). "Proteomic analysis of whey from bovine colostrum and mature milk." *Brazilian Archives of Biology and Technology*, *54*, 761-68.

[48] Golinelli, L., Mere Del Aguila, E., Paschoalin, V., Silva, J. & Conte Junior, C. (2014). "Functional aspect of colostrum and whey proteins in human milk." *Journal of Human Nutrition & Food Science*, *2*, 1035.

[49] Gomez-Gallego, C., Garcia-Mantrana, I., Salminen, S. & Collado, M. C. (2016). "The human milk microbiome and factors influencing its composition and activity." *Semin Fetal Neonatal Med*, 21, 400-05.

[50] Gómez-Gallego, C., Morales, J. M., Monleón, D., Du Toit, E., Kumar, H., Linderborg, K. M., Zhang, Y., Yang, B., Isolauri, E., Salminen, S. & Collado, M. C. (2018). "Human Breast Milk NMR Metabolomic Profile across Specific Geographical Locations and Its Association with the Milk Microbiota." *Nutrients*, *10*, 1355. https://doi.org/10.3390/nu10101355

[51] Gonzalez, R., Maldonado, A., Martín, V., Mandomando, I., Fumado, V. & Metzner, K. J. (2013). "Breast milk and gut microbiota in African mothers and infants from an area of high HIV prevalence." *PLoS One*, *8*, 80299.

[52] Gronlund, M. M., Gueimonde, M. & Laitinen, K. (2007). "Maternal breast-milk and intestinal bifidobacteria guide the compositional development of the Bifidobacterium microbiota in infants at risk of allergic disease." *Clin Exp Allergy*, *37*, 1764–72.

[53] Grönlund, M., Arvilommi, H., Kero, P., Lehtonen, O. & Isolauri, E. (2000). "Importance of intestinal colonisation in the maturation of humoral immunity in early infancy: A prospective follow up study of healthy infants aged 0–6 months." *Arch. Dis. Child. Fetal Neonatal Ed*, *83*, 186–92.

[54] Gueimonde, M., Sakata, S., Kalliom€aki, M., Isolauri, E., Benno, Y. & Salminen, S. (2006). "Effect of maternal consumption of Lactobacillus GG on transfer and establishment of fecal bifidobacterial microbiota in neonates." *Journal of Pediatr Gastroenterol Nutrition*, *42*, 166-70.

[55] Gupta, R., Jeevaratnam, K. & Fatima, A. (2018). "Lactic Acid Bacteria: Probiotic Characteristic, Selection Criteria, and its Role in Human Health (A Review)." *Journal of Emerging Technologies and Innovative Research*, *5*, 411-24.

[56] Handa and Sharma. (2016). "A Study on Bacteriocin Produced from a Novel Strain of Lactobacillus Crustorum F11 Isolated from Human Milk. *Proceedings of the Indian National Science Academy*, *82*(4).

[57] Handa, S. & Sharma, N. (2016). "A Study on Bacteriocin Produced from a Novel Strain of Lactobacillus crustorum F11 Isolated from Human Milk. *Proc Indian Natn Sci Acad*, *82*, No. 2, June 2016, pp.

[58] Hansen, L. A. & Soderstrom, T. (1981). "Human milk: Defence against infection". *Prog Clin Biol Res*, *61*, 147-59.

[59] Hanson, L. A., Korotkova, M., Lundin, S., Håversen, L., Silfverdal, S. A. & Mattsby-Baltzer, I. (2003). "The transfer of immunity from mother to child." *Annals of N Y Academic Sciences*, *987*, 199-206.

[60] Hanson, L. H., Sawicki, V., Lewis, A., Nuijens, J. H., Neville, M. C., Zhang, P. (2001). "Does human lactoferrin in the milk of transgenic mice deliver iron to suckling neonates?" *Advanced Experimental Medical Biology*, *501*, 233-39.

[61] Hartnett, D. J., Vaughan, A. & Sinderen, D. V. (2002). "Antimicrobial producing lactic acid bacteria isolated from Raw Barley and sorghum." *Journal of the Institute of Brewing*, *108*, 169-77.

[62] Heikkilä, M. & Saris, P. (2003). "Inhibition of Staphylococcus aureus by the commensal bacteria of human milk." *Journal of Applied Microbiology*, *95*, 471–78.

[63] Herbin, S., Mathiew, F., Brule, F., Branlant, C., Lefebvre, G. & Lebrihi, A. (1997). "Characteristics and genetic determinants of

bacteriocin activities produced by Cornobacterium piscicola CP5 isolated from cheese." *Current Microbiology*, *35*, 319-26.

[64] Hoashi, M., Meche, L., Mahal, L. K., Bakacs, E., Nardella, D. & Naftolin, F. (2015). "Human milk bacterial and glycosylation patterns differ by delivery mode." *Reprod Sci*, 2015, pii: 1933719115623645.

[65] Humphrey, B. D., Huang, N. & Klasing, K. C. (2002). "Rice expressing lactoferrin and lysozyme has antibiotic-like properties when fed to chicks." *Journal of Nutrition*, *132*, 1214-18.

[66] Isaacs, C. E. (2005). "Human milk inactivates pathogens individually, additively, and synergistically." *Journal of Nutrition*, *135*, 1286–88.

[67] Ismail, B. & Nampoothiri, K. M. (2010). "Production, purification and structural characterization of an exopolysaccharide produced by a probiotic *Lactobacillus plantarum* MTCC 9510." *Archives of Microbiology*, *192*, 1049-57.

[68] Ismail, I. H., Oppedisano, F. & Joseph, S. J. (2012). "Reduced gut microbial diversity in early life is associated with later development of eczema but not atopy in high-risk infants." *Pediatric Allergy Immunol*, *23*, 674–81.

[69] Italianer, M. F., Naninck, E. F. G., Roelants, J. A., Van der Horst, G. T. J., Reiss, I. K. M., Goudoever, J. B. V., Joosten, K. F. M., Chaves, I. & Vermeulen, M. J. (2020). "Circadian Variation in Human Milk Composition, a Systematic Review." *Nutrients*, *12*, 2328. https://doi.org/10.3390/nu12082328.

[70] Izadpanah, A. & Gallo, R. (2005). "Antimicrobial peptides." *Journal of the American Academy of Dermatology*, *52*, 381-90.

[71] Jabeen, N., Gul, H., Subhan, S. A., Hussain, M., Ajaz, M. & Rasool, S. A. (2009). "Biophysicochemical characterization of bacteriocins from indigenously isolated Agrobacterium radiobacter NA6." *Pakistan Journal of Botany*, *41*, 3227-37.

[72] Jeurink, P. V., Van Bergenhenegouwen, J. & Jimenez, E. (2013). "Human milk: a source of more life than we imagine." *Benef Microbes*, *4*, 17–30.

[73] Jimenez, E., Delgado, S. & Fernandez, L. (2008). "Assessment of the bacterial diversity of human colostrum and screening of staphylococcal and enterococcal populations for potential virulence factors." *Res Microbiol*, *159*, 595–601.

[74] Jimenez, E., Delgado, S. & Maldonado, A. (2008). "Staphylococcus epidermidis: a differential trait of the fecal microbiota of breast-fed infants." *BMC Microbiology*, *8*, doi:10.1186/1471-2180-8-143.

[75] Jost, T., Lacroix, C. & Braegger, C. (2013). "Assessment of bacterial diversity in breast milk using culture-dependent and culture-independent approaches." *Britain Journal of Nutrition*, *110*, 1253–62.

[76] Jost, T., Lacroix, C., Braegger, C. & Chassard, C. (2013). "Assessment of bacterial diversity in breast milk using culture-dependent and culture-independent approaches." *Britain Journal of Nutrition*, *14*, 1–10.

[77] Jost, T., Lacroix, C., Braegger, C. & Chassard, C. (2015). "Impact of human milk bacteria and oligosaccharides on neonatal gut microbiota establishment and gut health." *Nutrition Review*, *73*, 426-37.

[78] Jost, T., Lacroix, C., Braegger, C., Rochat, F. & Chassard, C. (2014). "Vertical mother-neonate transfer of maternal gut bacteria via breastfeeding." *Environmental Microbiology*, *16*, 2891-904.

[79] Jurkowski, M. & Błaszczyk, M. (2012). Charakterystyka fizjologiczno-biochemiczna bakterii fermentacji mlekowej. *Kosm. Probl. Nauk Biol.*, *3*, 493–504. [Physiological and biochemical characteristics of lactic acid bacteria]

[80] Kaingade, P., Somasundaram, I., Nikam, A., Behera, P. & Kulkarni, S. (2017). "Breast milk cell components and its beneficial effects on neonates: Need for breast milk cell banking." *Journal of Pediatric and Neonatal Individualized Medicine*, *6*, 1-12.

[81] Kaur, B. & Balgir, P. P. (2004). "Purification and characterization of an antimicrobial range of bacteriocin obtained from an isolate of *Pediococcus* species." *The Journal of Punjab Academy of Sciences*, *139*, 144.

[82] Kelleher, S. L., Chatterton, D., Nielsen, K. & Lönnerdal, B. (2003). "Glycomacropeptide and alpha-lactalbumin supplementation of infant formula affects growth and nutritional status in infant rhesus monkeys. *American Journal of Clinical Nutrition*, 77, 1261-68.

[83] Kennedy, R. S., Konok, G. P., Bounous, G., Baruchel, S. & Lee, T. D. (1995). "The use of a whey protein concentrates in the treatment of patients with metastatic carcinoma: a phase I-II clinical study." *Anticancer Research*, 15, 2643-49.

[84] Khalil, R., Elbahloul, Y., Fatima, D. F. & Omar, S. (2009). "Isolation and partial characterization of a bacteriocin produced by a newly isolated *Bacillus megaterium* 19 strain." *Pakistan Journal of Nutrition*, 8, 242-50.

[85] Khodayar-Pardo, P., Mira-Pascual, L. & Collado, M. C. (2014). "Impact of lactation stage, gestational age and mode of delivery on breast milk microbiota." *Journal of Perinatol*, 34, 599–605.

[86] Korhonen, H., Pihlanto-Leppala, A., Rantamaki, P. & Tupasela, T. (1998). "The functional and biological properties of whey proteins: prospects for the development of functional foods: a review." *Agricultural and Food Science in Finland*, 7, 283-96.

[87] Kormin, S. (1998). Isolation and screening of bacteriocin production LAB from tempeh. In: *Digitized by Web Admin team FSMB, LAB or their metabolic end products as bio-preservatives*, Radu S, Ali GRR and Jainudeen MR. (eds.) Putra University, Malaysia.

[88] Kunz, C., Rudloff, S., Baier, W., Klein, N. & Strobel, S. (2000). "Oligosaccharides in human milk: Structural, functional, and metabolic aspects." *Annual Review of Nutrition*, 20, 699-22.

[89] Lee-Huang, S., Maiorov, V., Huang, P. L., Ng, A., Lee, H. C., Chang, Y. T. & Kallenbach, N. (2005). "Structural and functional modeling of human lysozyme reveals a unique nonapeptide, HL9, with anti-HIV activity." *Biochemistry*, 44, 4648-55.

[90] Lönnerdal, B. & Atkinson, S. (1993). "Nitrogenous components of milk. Em: *Handbook of milk composition*. Robert G Jensen, editor. Academic Press, California., 1995, 351-368.

[91] Lönnerdal, B. & Glazier, C. (1985). "Calcium binding by alpha-lactalbumin in human milk and bovine milk." *Journal of Nutrition*, *115*, 1209-16.
[92] Lönnerdal, B. (1885). "Bioavailability of trace elements from human milk, cow's milk and infant formulas. Em: *Composition and physiological properties of human milk*. Schaub J, editor. Elsevier Science Publishers, Amsterdam, New York, Oxford. 1985; 3-14.
[93] Lönnerdal, B. (2004). "Human milk proteins: key components for the biological activity of human milk." *Advanced Experiment Medical Biology*, *554*, 11-25.
[94] Lönnerdal, B. (2006). "Recombinant human milk proteins." *Nestle Nutrition Workshop Ser Pediatr Program*, *58*, 207-15.
[95] Looijesteijn, P. J., Trapet, L., De vrie, E., Abee, T. & Hugenholtz, J. (2001). "Physiological function of exopolysaccharides produced by Lactococcus lactis". *International Journal of Food Microbiology*, *64*, 71-80.
[96] Looijesteijn, P. J., Van Casteren, W. H. C., Tuinier, C. H. L. & Doeswijk-Voragen, J. H. (2001). "Influence of different substrate limitations on the yield, composition, and molecular mass of exopolysaccharides produced by *Lactococcus lactis* subsp. cremoris in continuous cultures." *Journal of Applied Microbiology*, *89*, 116-22.
[97] Ma, J., Li, Z., Zhang, W., Zhang, C., Zhang, Y., Mei, H., Zhou, N., Wang, H., Wang, L. & Wu, D. (2020). "Comparison of gut microbiota in exclusively breast-fed and formula-fed babies: A study of 91 term infants." *Science Reporters*, *10*, 15792.
[98] Ma, J., Prince, A. L., Bader, D., Hu, M., Ganu, R. & Baquero, K. "High-fat maternal diet during pregnancy persistently alters the offspring microbiome in a primate model." *National Communication*, *5*, 3889.
[99] Makino, H., Kushiro, A. & Ishikawa, E. (2011). "Transmission of intestinal Bifidobacterium longum subsp. longum strains from mother to infant, determined by multilocus sequencing typing and

amplified fragment length polymorphism." *Applied Environmental Microbiology*, 77, 6788–93.

[100] Maldonado, J., Canabate, F. & Sempere, L. "Human milk probiotic *Lactobacillus fermentum* CECT5716 reduces the incidence of gastrointestinal and upper respiratory tract infections in infants." *Journal of Pediatr Gastroenterol Nutrition*, 54, 55–61.

[101] Maldonado-Barragan, A., Caballero-Guerrero, B. & Jimenez, E. (2009). "Enterocin C, a class IIb bacteriocin produced by *E. faecalis* C901, a strain isolated from human colostrum." *International Journal of Food Microbiology*, 133, 105–12.

[102] Mantis, N. J., Rol, N. & Corthésy, B. (2011). "Secretory IgA's complex roles in immunity and mucosal homeostasis in the gut". *Mucosal Immunol*, 6, 603-11.

[103] Marnila, P. & Korhonen, H. (2003). Colostrum. *ECM: Encyclopedia of Dairy Sciences*. Roginski H, editor. Academic Press, Londres., 1, 437-478.

[104] Martín, R., Heilig G Zoetendal, E., Smidt, H. & Rodríguez, J. (2007). "Diversity of the Lactobacillus group in breast milk and vagina of healthy women and potential role in the colonization of the infant gut." *Journal of Applied Microbiology*, 103, 2638–44.

[105] Martin, R., Langa, S. & Reviriego, C. (2004). "The commensal microflora of human milk: new perspectives for food bacteriotherapy and probiotics." *Trends Food Science Technology*, 15, 121–27.

[106] Martín, R., Olivares, M., Marín, M. L., Fernández, L., Xaus, J. & Rodríguez, J. M. (2005). "Probiotic potential of 3 lactobacilli strains isolated from breast milk." *J. Hum. Lact.*, 21, 8–17.

[107] Martín, V., Maldonado-Barragán, A., Moles, L., Rodriguez-Baños, M., Campo, R. D., Fernández, L., Rodríguez, J. M. & Jiménez, E. (2012). "Sharing of bacterial strains between breast milk and infant feces." *J. Human Lact. J. Int. Lact. Consult. Assoc.*, 28, 36–44.

[108] Martinez, R. C. R. & De Martinis, E, C, P. (2006). "Effect of *Leuconosoc mesenteroides* ll bacteriocin in the multiplication control of *Listeria monocytogenes*." *Ciencia Technologia de-Alimentos*, 26, 52-55.

[109] Martínez-García, R. M. (2002). *Nutritional status of a group of pregnant Watershed. Influence on the composition of breast milk.* Doctoral Thesis in Nutrition. Department of Nutrition and Food Science I, School of Pharmacy, Complutense University of Madrid, Cuenca, Spain., 2002.

[110] Mastroimarino, P., Capobianco, D., Miccheli, A., Pratico, G., Campagna, G. & Laforgia, N. (2015). "Administration of a multistrain probiotic product (VSL#3) to women in the perinatal period differentially affects breast milk beneficial microbiota in relation to mode of delivery." *Pharmacol Res*, *95*, 63-70.

[111] Mata Llrrutia, L. J. & Urrutia, J. J. (1971). "Intestinal colonization of breast-fed children in a rural area of low socioeconomic level." *Annals of the New York Academy of Sciences*, *176*, 93.

[112] Mataix, J. & Hernandez, M. (2002). Infant. Em: Nutrition and Food Humana. II. Physiological and pathological situations. Mataix J, editor. *HERGON, Madrid.*, 2002, 835.

[113] Matsumiva, Y., Kato. N., Watanabe, K. & Kato, H. (2002). "Molecular epidemiological study of vertical transmission of vaginal Lactobacillus species from mothers to newborn infants in Japanese, by arbitrarily primed polymerase chain reaction." *Journal of Infection Chemother*, *8*, 43–49.

[114] McIntosh, G. H., Regester, G. O., Le Leu, R. K., Royle, P. J. & Smithers, G. W. (1995). "Dairy proteins protect against dimethylhydrazine-induced intestinal cancers in rats." *Journal of Nutrition*, *125*, 809-16.

[115] Meera, N. S. & Charitha, M. (2012). *Partial characterization and optimization of parameters for Bacteriocin production by Probiotic Lactic acid bacteria.*, 2.

[116] Michael, J. G., Ringenback, R. & Hottenstein, S. (1971). "The antimicrobial activity of human colostral antibody in the newborn." *Journal of Infectious Diseases*, *124*, 445.

[117] Mojgani, N., Sabiri, G., Ashtiani, M. P. & Torshizi, M. A. K. (2009). "Characterization of bacteriocins produced by *Lactobacillus brevis*

NM 24 and L. fermentum NM 332 isolated from green olives in Iran." *The International Journal of Microbiology*, 6.

[118] Motta, A. S. & Brandelli, A. (2003). "Influence of growth conditions on bacteriocin production by Brevibacterium linens." *Applied and Microbial Biotechnology*, 62, 163-67.

[119] Nallala, V., Sadishkumar, V. & Jeevaratnam, K. (2017). "Molecular characterization of antimicrobial *Lactobacillus* isolates and evaluation of their probiotic characteristics *in vitro* for use in poultry." *Food Biotechnology*, 31, 20–41.

[120] Nishimura, R. Y., Barbieiri, P., Castro, G. S., Jordao, Jr. A. A., Perdona Gda. S. & Sartorelli, D. S. (2014). "Dietary polyunsaturated fatty acid intake during late pregnancy affects fatty acid composition of mature breast milk. *Nutrition*, 30, 685-89.

[121] Noonpakdee, W., Jumriangrit, P., Wittayakom, K., Zendo, J., Nakayama, Sonomoto K. & Panyim, S. (2009). "Two-peptide bacteriocin from Lactobacillus plantarum PMU 33 strain isolated from som-fak, a Thai low salt fermented fish product." *Asia Pacific Journal of Molecular Biology and Biotechnology*, 17, 19-25.

[122] Nwodo Uchechukwu, U., Ezekiel, G. & Anthony, I. O. (2012). "Bacterial Exopolysaccharides: Functionality and Prospects." *International Journal of Molecular Science*, 13, 14002-15.

[123] Olivares, M., Albrecht, S., De Palma, G., Ferrer, M. D., Castillejo, G., Schols, H. A. & Sanz, Y. "Human milk composition differs in healthy mothers and mothers with celiac disease." *European Journal of Nutrition*, 54, 119-28.

[124] Olver, W. J., Bond, D. W. & Boswell, T. C. (2003). "Neonatal group B streptococcal disease associated with infected breast milk. *Arch Dis Child Fetal Neonatal Ed.*, 2000, 83, F48–F49.

[125] Palmeira, P. & Carneiro-Sampaio, M. (2016). "Immunology of breast milk". *Rev Assoc Med Bras.*, 6, 584-593.

[126] Panagos, P. G., Vishwanathan, R., Penfield-Cyr, A., Matthan, N. R., Shivappa, N. & Wirth, M. D. (2016). "Breastmilk from obese mothers has pro-inflammatory properties and decreased

neuroprotective factors." *Journal of Perinatol* http://dx.doi.org/ 10.1038/jp.2015.199.

[127] Parvez, K. A., Malik, S., Kang, A. & Kim, H. Y. (2006). "Probiotics and their fermented food products are beneficial for health." *Journal of Applied Microbiology*, 1171-85.

[128] Parvez, S., Malik, K. A., Kang, S. A. & Kim, H. Y. (2006). "Probiotics and their fermented food products are beneficial for health." *Journal of Applied Microbiology*, *34*, 1364-5072.

[129] Pasqua, A. Q., Giuseppina, P., Liliana, C., Paola, L., Maria, A. G. & Antonio, V. (2021). "The Revolution of Breast Milk, The Multiple Role of Human Milk Banking between Evidence and Experience—A Narrative Review." *International Journal of Pediatrics*, vol. 2021, Article ID 6682516, 11 pages, 2021. https://doi.org/10. 1155/2021/6682516.

[130] Patel, A. R., Lindstrom, C., Patel, A., Prajapati, J. & Holst, O. (2012). "Screening and isolation of exopolysaccharide producing lactic acid bacteria from vegetables and indigenous fermented foods of Gujarat, India." *International Journal of Fermented Foods*, *1*, 87-101.

[131] Perez, P. F., Dore, J. & Leclerc, M. (2007). "Bacterial imprinting of the neonatal immune system: lessons from maternal cells?." *Pediatrics*, *119*, 724–32.

[132] Pusceddu, M. M., Aidy, E. S., Crispie, F., O'Sullivan, O., Cotter, P. & Stanton, C. (2015). "N-3 polyunsaturated fatty acids (PUFAs) reverse the impact of early-life stress on the gut microbiota." *PLoS One*, *10*, 0139721.

[133] Quinn, E. A., Largado, F., Power, M. & Kuzawa, C. W. (2012). "Predictors of breast milk macronutrient composition in Filipino mothers." *American Journal of Human Biology*, *24*, 533-40.

[134] Ramsay, D. T., Mitoulas, L. R. & Kent, J. C. (2005). "The use of ultrasound to characterize milk ejection in women using an electric breast pump." *Journal of Human Lactation*, *21*, 421–28.

[135] Rautava, S., Kainonen, E., Salminen, S. & Isolauri, E. (2012). "Maternal probiotic supplementation during pregnancy and breast-

feeding reduces the risk of eczema in the infant." *Journal of Allergy Clinical Immunology*, *130*, 1355-60.
[136] Reid, G., Kumar, H., Khan, A. I., Rautava, S., Tobin, J. & Salminen, S. (2016). "The case in favour of probiotics before, during and after pregnancy: insights from the first 1,500 days." *Beneficial Microbes*, *3*, 1-10.
[137] Rinne, M. M., Gueimonde, M., Kalliomaki, M., Hoppu, U., Salminen, S. J. & Isolauri, E. "Similar bifidogenic effects of prebiotic-supplemented partially hydrolyzed infant formula and breastfeeding on infant gut microbiota." *FEMS Immunology and Medical Microbiology*, *43*, 59–65.
[138] Rivero Urgell, M., Santamaría Orleans, A. & Rodríguez-Palmero Seuma, M. (2020). "The importance of functional ingredients in pediatric milk formulas and cereals". *Nutritional Hospital.*, *20*, 135-46.
[139] Samuel, T. M., Zhou, Q., Giuffrida, F., Munblit, D., Verhasselt, V. & Thakkar, S. K. (2020). "Nutritional and Non-nutritional Composition of Human Milk Is Modulated by Maternal, Infant, and Methodological Factor." *Frontiers in Nutrition*, *7*, 172.
[140] Sánchez, L. M., Martin, S. C. & Gómez-de-Orgaz, C. S. (2020). "Human milk bank and personalized nutrition in the NICU: a narrative review." *European Journal of Pediatr*, (2020). https://doi.org/10.1007/s00431-020-03887-y
[141] Sanchez-Medina, I., Gerwig, G. J., Urshev, Z. L. & Kamerlinga, J. P. (2007). "Structural determination of a neutral exopolysaccharide produced by *Lactobacillus delbrueckii* ssp. bulgaricus LBB.B332." *Carbohydrate Research*, *342*, 2735–44.
[142] Savadogo, A., Ouattara, C. A. T., Bassole, H. N. I. & Traore, A. S. (2004). "Antimicrobial activities of lactic acid bacteria strains isolated from burkina faso fermented milk." *Pakistan Journal of Nutrition*, *3*, 174-179.
[143] Savadogo, A., Ouattara, C. A. T., Savadogo, P. W., Barro, N., Ouattara, A. S. & Traore, A. S. (2004). "Identification of exopolysaccharides-producing lactic acid bacteria from Burkina

Faso fermented milk samples." *African Journal of Biotechnology*, *3*, 189-94.

[144] Serrano-Niño, J. C., Solís-Pacheco, J. R., Gutierrez-Padilla, J. A., Cobián-García, A., Cavazos-Garduño, A., González-Reynoso, O. & Aguilar-Uscanga, B. R. (2016). "Isolation and Identification of Lactic Acid Bacteria from Human Milk with Potential Probiotic Role." *Journal of Food and Nutrition Research*, *4*, 170–77.

[145] Sharma, K., Sharma, N. & Handa, S. (2020). "Purification and characterization of novel exopolysaccharides produced from *Lactobacillus paraplantarum* KM1 isolated from human milk and its cytotoxicity." *J Genet Eng Biotechnol*, *18*, 56.

[146] Sharma, N. & Gautam, N. (2008). "Antibacterial activity and characterization of bacteriocin of *Bacillus mycoides* isolated from whey." *Indian Journal of Biotechnology*, *7*, 117-21.

[147] Sharma, N., Kapoor, G. & Neopaney, B. (2006). "Characterization of a new bacteriocin produced from a novel isolated strain of *Bacillus lentus* NG 121." *Antonie van Leeuwenhock*, *89*, 337-43.

[148] Sharma, N., Kapoor, R., Gautam, N. & Kumari, R. (2011). "Purification and characterization of bacteriocin produced by Lactobacillus sp. A75 isolated from fermented chunks of Phaseolusradiata." *Food Technology and Biotechnology*, *49*, 169-76.

[149] Shin, K., Hayasawa, H. & Lönnerdal, B. (2001). "Purification and quantification of lactoperoxidase in human milk with use of immune adsorbents with antibodies against recombinant human lactoperoxidase." *American Journal of Clinical Nutrition*, *73*, 984-89.

[150] Shouny, W. E., Kumar, A., Shanshury, A. E. E. & Ragy, S. (2012). "Production of plantaricin by L. plantarum S R 18." *Journal of Microbiology, Biotechnology and Food Sciences*, *1*, 1488-1504.

[151] Soto, A., Martín, V., Jiménez, E., Mader, I., Rodríguez, J. M. & Fernández, L. (2014). "Lactobacilli and bifidobacterial in human breast milk, Influence of antibiotherapy and other host and clinical factors." *Journal of Pediatric Gastroenterol Nutrition*, pp. 59.

[152] Spatz, D. L., Davanzo, R., Müller, J. A., Powell, R., Rigourd, V., Yates, A., Geddes, D. T., Van Goudoever, J. B. & Bode, L. (2021). "Promoting and Protecting Human Milk and Breastfeeding in a COVID-19 World." *Front. Pediatr.*, *8*, 633700.

[153] Spatz, D. L. (2020). "Using the coronavirus pandemic as an opportunity to address the use of human milk and breastfeeding as lifesaving medical interventions." *Journal Obstet Gynecol Neonatal Nurs*, *49*, 225–36.

[154] Sylvetsky, A. C., Gardner, A. L., Bauman, V., Blau, J. E., Garraffo, H. M. & Walter, P. J. (2015). "Nonnutritive sweeteners in breast milk." *Journal of Toxicology Environment Health*, *78*, 1029-32.

[155] Taipale, T. J., Isolauri, E., Jokela, J. T. & S€oderling, E. M. (2016 Jan). Bifidobacterium animalis subsp. lactis BB-12 in reducing the risk of infections in early childhood. *Pediatr Res*, *79*(1-1), 65e9.

[156] Takako, H., Mizue, M., Izumi, H., Chie, O., Harue, T. & Uchida, M. (2020). "Improving human milk and breastfeeding rates in a perinatal hospital in Japan: a quality improvement project." *Breastfeed Med*, *15*, 538–45.

[157] Tassovatz, K., El-Hawary, M. F. S., Soliman, A. A. & Nosseir, S. A. (1981). "Biochemical studies on the effect of breast and artificial feeding in newborn Egyptian infants I. Serum proteins and immunoglobulins in 1–4-day-old newborns." *Z Ernährungswiss*, *20*, 283–90.

[158] Todorov, S. D. & Dicks, L. M. T. (2005). "Effect of growth medium on bacteriocin production by *Lactobacillus plantarum* ST194BZ., a strain isolated from boza." *Food Technology and Biotechnology*, *43*, 165-73.

[159] Todorov, S. D., Velho, V. M. & Dicks, L. M. T. (2004). "Isolation and partial characterization of bacteriocins produced by four lactic acid bacteria isolated from traditional South African Beer." *Electronic Journal of Environment*, *72*, 559-64.

[160] Tolinacki, M., Kojic, M., Lozo, J., Vidojevic, A., Topisirovic, L. & Fira, D. (2010). "Characterization of bacteriocin producing strain

Lactobacillus paracasei subsp. paracasei BGUB9." *Biomedical and Life Sciences*, *62*, 889-99.

[161] Tsuda, H. & Miyamoto, T. (2010). "Production of exopolysaccharide by *Lactobacillus plantarum* and the prebiotic activity of the exopolysaccharide." *Food Science and Technology Research*, *16*, 87-92.

[162] Turin, C. G. & Ochoa, T. J. (2014). "The role of maternal breast milk in preventing infantile diarrhea in the developing world." *Current Trop Med Rep.*, *1*, 97-105.

[163] Urba´ nska, M. & Szajewska, H. (2014). "The efficacy of Lactobacillus reuteri DSM 17938 in infants and children: A review of the current evidence." *European Journal of Pediatrics*, *173*, 1327–1337.

[164] Urbaniak, C., Angelini, M., Gloor, G. B. & Reid, G. (2016). "Human milk microbiota profiles in relation to birthing method, gestation and infant gender." *Microbiome*, *4*, 1.

[165] Urbaniak, C., McMillan, A., Angelini, M., Gloor, G. B., Sumarah, M. & Burton, J. P. "Effect of chemotherapy on the microbiota and metabolome of human milk, a case report." *Microbiome*, *2*, 24.

[166] Ushida, T., Oda, T., Sato, K. & Kawakami, H. (2006). "Availability of lactoferrin as a natural solubilizer of iron for food products." *International Dairy Journal*, *16*, 95-101.

[167] Ustundag, B., Yilmaz, E., Dogan, Y., Akarsu, S., Canatan, H. & Halifeoglu, I. (2005). "Levels of Cytokines (IL-1β, IL-2, IL-6, IL-8, TNF-α) and Trace Elements (Zn, Cu) in Breast Milk from Mothers of Preterm and Term Infants." *Mediators Inflamm*, 2005, 331-36.

[168] Vescovo, M., Torriani, S., Orsi, C., Macchiorolo, F. & Solari, G. (1996). "Application of antimicrobial producing lactic acid bacteria to control pathogens in ready to use vegetables." *Journal of Applied Biotechnology*, *81*, 113-19.

[169] Vincent, S. J. F., Faber, E. J., Neeser, J., Stingele, F. & Kamerling, J. P. (2001). "Structure and properties of the exopolysaccharide produced by *Streptococcus macedonicus* Sc136." *Glycobiology*, *11*, 131-39.

[170] Wada, T., Noda, M., Kashiwabara, F., Jeon, H. J., Shirakawa, A., Yabu, H., Matoba, Y., Kumagai, T. & Sugiyama, M. (2009). "Characterization of four plasmids harboured in a Lactobacillus brevis strain encoding a novel bacteriocin, brevicin 925A, and construction of a shuttle vector for lactic acid bacteria and Escherichia coli." *Microbiology*, *155*, 1726-37.

[171] Wagner, C. L., Taylor, S. N. & Johnson, D. (2008). "Host factors in amniotic fluid and breast milk that contribute to gut maturation." *Clinical Review of Allergy Immunology*, *34*, 191–204.

[172] Wakabayashi, H., Yamauchi, K. & Takase, M. (2006). "Lactoferrin research, technology and applications." *International Dairy Journal*, *16*, 1241-51.

[173] Walker, A. (2010). "Breast milk as the gold standard for protective nutrients." *Journal of Pediatr*, *156*, 3-7.

[174] Wang, S. C., Chang, C. K., Chan, S. C., Shieh, J. S., Chiu, C. K. & Duh, P. D. (2014). "Effects of lactic acid bacteria isolated from fermented mustard on lowering cholesterol." *Asian Pacific Journal of Tropical Biomedicine*, *4*, 523-28.

[175] West, P. A., Hewitt, J. H. & Murphy, O. M. (1979). "Influence of methods of collection and storage on the bacteriology of human-milk." *Journal of Applied Bacteriology*, *46*, 269-77.

[176] Witt, A., Mason, M. J., Burgess, K., Flocke, S. & Zyzanski, S. A. (2014). "Case control study of bacterial species and colony count in milk of breastfeeding women with chronic pain." *Breastfeed Med*, *9*, 29-34.

[177] Wu, S., Grimm, R., German, J. B. & Lebrilla, C. B. (2011). "Annotation and structural analysis of sialylated human milk oligosaccharides." *Journal of Proteome Research*, *10*, 856-68.

ABOUT THE AUTHOR

Nivedita Sharma is working as Professor and Head in the Department of Basic Sciences, Dr Y S Parmar university of Horticulture and Forestry Nauni, Solan (H.P.) –India. She has 25 years of teaching and research experience in Microbiology discipline. She is actively involved in the research area of probiotics, functional foods, neutraceutical and biofuels. She has guided many MSc and PhD students. She has published more than 200 publications in International and National journals and Proceedings. She is an active organizing committee member for seminars, conferences, workshops of the department and university. She is the recipient of many National and International awards. She is the member of many prestigious societies.

E-mail id: niveditashaarma@yahoo.co.in

In: Human Milk
Editor: John I. Cole

ISBN: 978-1-53619-713-6
© 2021 Nova Science Publishers, Inc.

Chapter 2

FATTY ACID SUPPLY OF HUMAN MILK AND ITS POSSIBLE HEALTH EFFECTS

Éva Szabó[*]
Department of Biochemistry and Medical Chemistry,
University of Pécs, Medical School, Pécs, Hungary

ABSTRACT

Human milk is the optimal choice for infant nutrition, with exclusive breastfeeding for the first six months of life and introduction of complementary feeding should also start along with continued breastfeeding. Long chain polyunsaturated fatty acids (LCPUFAs) play an important role in building up membranes and they are also precursors of certain second messengers. The two most important LCPUFAs, the n-3 docosahexaenoic acid (DHA) and n-6 arachidonic acid (AA) are the predominant fatty acids in human brain and are key factors in the visual- and neurodevelopment during the third trimester and in the first months of life. For breastfed infants the exclusive source of these fatty acids is human milk which can be influenced by maternal diet. Several studies have found a difference between breastmilk of mothers who gave birth to term newborn and breastmilk of mothers who gave birth of preterm infants. Preterm neonates have only limited body stores of LCPUFAs but

---

[*] Corresponding Author's E-mail: szabo.eva.dr@pte.hu.

their dietary requirement for these fatty acids is increased due to their rapidly developing tissues. Therefore, the fatty acid composition of breast milk of mothers who gave birth to preterm or term, mature babies is different in order to cover their different dietary needs. The fatty acid composition of breastmilk is influenced not only by the gestational age at birth, but also by the time of sampling. Several studies have found that fatty acid composition of colostrum is different from that of mature milk. These results suggest that fatty acid profile of breastmilk varies throughout lactation adapting to the nutritional needs of the developing infant. The main sources of fatty acids in human milk are maternal diet as well as maternal stores. Therefore, some of the differences found in the fatty acid composition of milk samples can be attributed to different eating habits in different countries. This chapter gives an overview of the fatty acid composition of breastmilk in mothers who gave birth to newborn of different gestational ages, as well as during lactation or as a result of diet.

Keywords: arachidonic acid, colostrum, diet, docosahexaenoic acid, fat content, fatty acid, gestational age, human milk, lactation, mature milk, nationality, preterm, term, transitional milk

INTRODUCTION

Human milk is the best form of nutrition for infants containing all the important nutrients for healthy growing. WHO (WHO 2018) recommends 6 months exclusive breastfeeding followed by continued breastfeeding as complementary foods are introduced. Breastfeeding can be continued until 1 year of age or longer according to maternal and infant's needs or desires. The exclusive breastfeeding for 6 months is by other committees, like the American Academy of Pediatrics (AAP) (AAP 2012), the Academy of Nutrition and Dietetics (Lessen and Kavanagh 2015) and the European Society for Pediatric Gastroenterology, Hepatology, and Nutrition (ESPGHAN) (ESPGHAN et al. 2009) also recommended.

Human milk provides energy and macromolecules for growing. A major source of energy in milk are lipids accounting for about 44% of energy intake. When we calculate with 6 months exclusively breastfeeding this results in about 3.9 kg lipids and 35.000 kcal supplied to fully breastfed infants (Koletzko 2016). According to a recent systematic review (Perrin et al. 2020), donor human milk (a pooled milk sample of mothers with different lactation stage, with no interindividual diversity) contains an average of 49.3 - 69.3 kcal/100 mL energy and 1.8 - 4.1 g/100 mL fat (**Table 1**). Among other important nutrients, term human milk contains about 1.8 - 3.6 g/100 mL fat (Gidrewicz and Fenton 2014), mainly in form of triacylglycerols (TG), which accounts for about 98% (Bitman et al. 1983, Harzer et al. 1983). The other important lipid fractions are phospholipids (PLs) and cholesterol. These components form the so called milk fat globules, containing non-polar lipids, mainly TGs and some mono- and diacylglycerols in the centre surrounded by a lipid bilayer that contain different phospholipids, cholesterol, complex lipids and bioactive peptides.

The quality and quantity of fat content in human milk can be influenced by numerous factors, like the course of lactation (Koletzko 2016, Gidrewicz and Fenton 2014, Bahreynian, Feizi, and Kelishadi 2020, Floris et al. 2020, Berenhauser et al. 2012, Minda et al. 2004, Kovács et al. 2005, Mihalyi et al. 2015), gestational week at delivery (Floris et al. 2020, Berenhauser et al. 2012, Kovács et al. 2005, Molto-Puigmarti et al. 2011), maternal diet (Karcz and Krolak-Olejnik 2020, Liu et al. 2016, Aumeistere et al. 2019, Siziba, Lorenz, et al. 2020) and nationality of mothers (Fu et al. 2016, Bahreynian, Feizi, and Kelishadi 2020, Rueda et al. 1998, Sinanoglou et al. 2017). The time of sampling also influences fat content of human milk as hindmilk contains more fat than foremilk (Budowski et al. 1994, Demmelmair and Koletzko 2018, Karatas et al. 2011, Siziba, Chimhashu, et al. 2020) and circadian variation also exists with highest fat content in the evening (Italianer et al. 2020, Harzer et al. 1983, Paulaviciene et al. 2020, Jackson et al. 1988, Demmelmair and Koletzko 2018). According to some recent data infants' sex might also affect the fat content (de Fluiter et al. 2020, Hosseini et al. 2020, Fischer Fumeaux et al. 2019) but others didn't confirm this (Mangel et al. 2020, Bzikowska-Jura et al. 2020, Sinanoglou et al. 2017). These factors may explain the inter- and intraindividual variation affecting milk fat concentration.

Table 1. Average nutrient contents of donor human milk (data modified according to Perrin et al., 2020)

	Average values (number of studies reviewed)
Energy (kcal/100 mL)	49.3 - 69.3 (n = 6)
Total carbohydrate (g/dL)	6.5 - 7.4 (n = 4)
Total fat (g/dL)	1.8 - 4.1 (n = 10)
Total protein (g/dL)	0.8 - 3.2 (n = 10)

FATTY ACIDS

Fatty acids play an important role in the human organism in building up cell membranes (mainly as PLs), storing energy in adipocytes (mainly in the form of TGs) and as precursors of several lipid mediators such as leukotrienes, prostaglandins and thromboxanes. The physiologically most important fatty acids are the n-3 and n-6 polyunsaturated fatty acids (PUFAs) containing at least two double bonds in cis configuration. Essential fatty acids (EFAs) cannot be synthesized by the human body so we have to take them with our food. These EFAs occur primarily in plant foods, like vegetable oils and oily seeds. Flaxseed oil is very rich in the essential n-3 α-linolenic acid (C18:3n-3, ALA) but chia seed, walnut and hempseed oil also contain significant amounts of it (Table 2). The essential n-6 fatty acid, linoleic acid (C18:2n-6, LA) is found in the highest proportion in safflower and grape seed oil, but the most relevant sources in human nutrition are sunflower and corn oils (Table 2). Although food of animal origin contains negligible amounts of ALA, they are rich sources of LA, mainly poultry meat.

From the EFAs longer chain fatty acids can be synthesized. The simplified metabolism of fatty acids is shown in Figure 1. The most important long chain metabolite of the essential LA is arachidonic acid (C20:4n-6, AA) and the essential n-3 ALA is the precursor of eicosapentaenoic acid (C20:5n-3, EPA) docosapentaenoic acid (C22:5n-3, DPA) and docosahexaenoic acid (C22:6n-3, DHA). The elongations occur quite fast, while desaturation is much slower. From the two desaturase enzymes the Δ-6 desaturation is the rate limiting step. The metabolism of

DHA is more complex: after elongation and desaturation peroxisomal β-oxidation is also needed. There is a competition between n-3 and n-6 fatty acids as they use the same enzymes during their metabolism. Although, both Δ-5 and Δ-6 desaturases prefer n-3 to n-6 fatty acids (Simopoulos 2010), high dietary LA intake can decrease the conversion of ALA to DHA. In some animal studies higher brain DHA was detected when feeding lower LA diet compared to high LA diet, while others reported no effect (Alashmali, Hopperton, and Bazinet 2016). The effect of lowering AA intake in the early postnatal life also showed contradictory results, some animal studies reported higher brain DHA while others showed no effect (Alashmali, Hopperton, and Bazinet 2016).

The ability of humans to convert essential fatty acids to their longer chain metabolites, namely AA, EPA, DPA and DHA is an important nutritional question, but previous studies have shown that dietary ALA-supplementation can result in only a small increase in EPA levels, but has no effect on DHA levels (Brenna et al. 2009). Isotopic tracer studies have found that the conversion of ALA to DHA is about 1% in infants and even more limited in adults (Brenna et al. 2009). A tracer study in lactating women has shown that the direct conversion of AA from LA didn't exceed 3.2%, while about 3-25% of C20:3n-6 was directly converted from dietary LA (Demmelmair et al. 1998). In healthy Mexican lactating women even lower (1.1%) conversion rate from dietary LA to AA was found in secreted milk samples (Del Prado et al. 2001). According to another tracer study using ^{13}C ALA about 10.9% of EPA, 4.4% of DPA and only 1.1% of DHA in human milk seemed to be directly derived from dietary ALA (Demmelmair et al. 2016).

As the conversion of LCPUFAs from their preformed EFAs is limited in the human organism, in conditions with increased demand (pregnancy, lactation, perinatal development), these fatty acids must be covered with our diet. The most important dietary sources of long-chain metabolites are animals. The main AA sources are meat and offal of herbivorous animals, like beef heart, liver, while EPA and DHA can be found in predatory or small fatty sea fishes, like salmon, tuna, mackerel, sardine and herring.

Table 2. Some selected vegetable oils rich in essential fatty acids

g fatty acid in 100 g oil	LA (C18:2n-6)	ALA (C18:3n-3)
Chia seed	5.9	**22.8**
Corn oil	54.5	1.2
Cottonseed oil	51.5	0.2
Flaxseed oil	14.3	**53.4**
Grape seed oil	**69.6**	0.1
Hempseed oil	27.4	**8.7**
Poppy seed oil	62.4	-
Safflower oil	74.6	-
Sesame oil	41.3	0.3
Soy oil (low LA)	56.0	3.0
Soybean oil	51.0	**6.8**
Sunflower oil (LA)	**65.7**	-
Walnut oil	52.9	**10.4**
Wheat germ oil	54.8	6.9

(Source: USDA FoodData Central, https://fdc.nal.usda.gov/fdc-app.html#/)

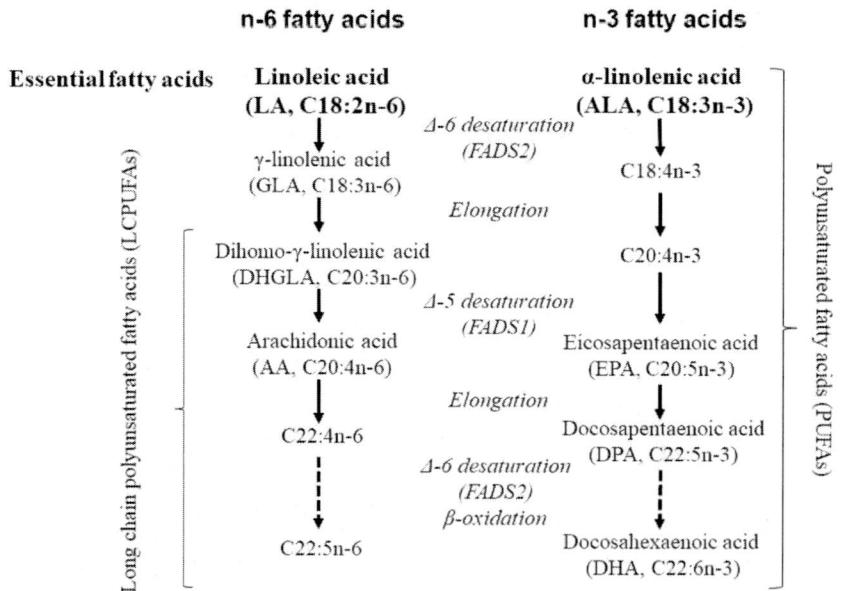

Figure 1. Metabolism of fatty acids.

In a study investigating the relationship between human milk fatty acids and average national cognitive performance measured with the PISA test significant positive correlations were found between human milk DHA content and all three investigated scores (math, reading, science), while higher milk LA was a significant negative predictor for scores in reading and science. This result can suggest that high dietary LA can suppress desaturase enzymes and possibly compete with DHA for incorporation into plasma PLs (Lassek and Gaulin 2014).

When supplementing lactating women with ALA, after 2 weeks breast milk ALA composition significantly increased while its metabolites, EPA, DPA and DHA didn't differ from baseline significantly (Francois et al. 2003) suggesting a limited conversion of ALA to n-3 LCPUFAs. In contrast, DHA supplementation during lactation increased both absolute and relative breast milk DHA content after 6 weeks (Sherry, Oliver, and Marriage 2015). A dose of 200 mg/day DHA also improved DHA status in mothers and their infants in a European study (Bergmann et al. 2008).

FATTY ACIDS AND NEURODEVELOPMENT

PUFAs play a key role in the infant growth and neurodevelopment. Due to rapid brain growth and phospholipid incorporation into the cerebral cortex in the perinatal period, there is a high demand of LCPUFAs, which has to be covered by nutrition. Brain weight increases about 4-5-fold during the last trimester and the accretion of AA and DHA increases from the 30th gestational week until birth (Clandinin et al. 1980b). After birth, postnatal brain growth is still increasing, but in a slower rate than during intrauterine life. Compared to intrauterine rates, the absolute accretion of AA exceeded the antenatal levels and was greatly increased from the 4th week to the 12th week after birth (Clandinin et al. 1980a).

In this PUFA accumulation several mechanisms could stand (Qi, Hall, and Deckelbaum 2002). Nonesterified fatty acids can be transported by a passive diffusion through the phospholipid bilayer or by facilitated transport through different transport proteins: fatty acid transport proteins (FATPs) and fatty acid binding proteins (FABPs). FATPs are membrane-bound proteins that facilitate tissue uptake of PUFAs, while FABPs are cytosolic proteins contributing not only FA uptake of the cell but intracellular FA trafficking also (Qi, Hall, and Deckelbaum 2002). The DHA accumulation in brain compared to plasma levels (Lacombe, Chouinard-Watkins, and Bazinet 2018) in many PL classes suggest a selective transport of LCPUFAs into the brain.

In a former study (Farquharson et al. 1992), significantly higher DHA values were found in the cerebral cortex of breast-fed infants than in formula fed. In those years, formulas didn't contain LCPUFAs (namely AA and DHA), only essential fatty acids, LA and ALA. This study also has shown that during rapid brain development the synthesis of DHA from its precursor, ALA is not sufficient to supply the brain with adequate amounts. In the perinatal period increased DHA incorporation into the forebrain has been observed in two PL fractions, phosphatidylcholine (PC) and phosphatidylethanolamine (PE) (Martinez and Mougan 1998), while concentration of AA remained quite stable. The DHA content of cerebral cortex is not only influenced by the nutrition (breastfed infants have higher content) but the accumulation of DHA also depends on the length of breastfeeding (Makrides et al. 1994). This could also have been the reason for measuring better visual acuity in breastfed infants at 5 months of age compared to formula-fed infants (Makrides et al. 1993).

Breastfed infants also have higher DHA concentration in their erythrocyte membranes (Makrides et al. 1994). Several studies have found an improvement in both cognitive and visual outcome of infants with higher blood DHA, while other found either no or even negative correlation (Hoffman, Boettcher, and Diersen-Schade 2009). Taking into account that DHA plays a key role in the perinatal neurodevelopment, the European Commission (Koletzko et al. 2008), the European Food Safety Authority (EFSA Panel on Dietetic Products 2010) and the Early Nutrition Academy (Koletzko et al. 2014) recommended at least 200 mg/day average intake of DHA to lactating women.

Based on the above mentioned roles in the early postnatal neurodevelopment, the question was raised, whether LCPUFA supplementation (mainly DHA) to lactating mothers can improve early infant development resulting in better language, psychomotor development or better visual acuity. A Cochrane review didn't find strong evidence that maternal LCPUFA supplementation improves child growth and neurodevelopment, however, the number of studies in the subgroup analyses was low, so further studies are needed (Delgado-Noguera et al. 2015). A recent review (Gawlik et al. 2020) also corroborated that early DHA supplementation has no effect of improved language development (9 months-10 years).

FATTY ACID CHANGES DURING LACTATION

Breast milk composition seems to be quite stable between 2-12 weeks, but there are some initial fluctuations as the milk changes from colostrum to mature human milk (Gidrewicz and Fenton 2014). Compared to colostrum, mature milk has a 16% increase in energy (54 vs. 63 kcal/dL) and a 93% increase in fat content (1.8 vs. 3.4 g/dL) in term milk (Gidrewicz and Fenton 2014). The ratios of saturated to monounsaturated and polyunsaturated fatty acids as well as the major fatty acids seem to remain quite stable over time, but it is suggested, that physiologically important minor components can change in the course of lactation. On the basis of a pooled data analysis including 55 studies worldwide the two most important LCPUFAs, namely AA and DHA seem to decrease over time in term human milk (Floris et al. 2020).

When we look on individual studies about fatty acid changes over lactation, the results might be more diverse (Table 3). Some authors found no change in AA and DHA values at different time points in mature milk (Barreiro et al. 2018, Grote et al. 2016) or between transitional an mature milk samples (Khor et al. 2020). Rueda at al. (Rueda et al. 1998) found no change in Spanish, but significantly decreased AA values in Panamanian lactating women. Other authors found significantly decreased DHA values in the course of lactation with quite stable AA values (Aydin et al. 2014, Barrera et al. 2018, Deng et al. 2018, Li et al. 2009). In contrast, others had opposite results: AA values significantly decreased while DHA values remained quite stable (Antonakou et al. 2013, Bobinski et al. 2013, Ribeiro et al. 2008, Sanchez-Hernandez et al. 2019). But in most studies both AA and DHA values became significantly lower over lactation (Berenhauser et al. 2012, Boersma et al. 1991, Decsi 2000, de Souza Santos da Costa et al. 2016, Genzel-Boroviczény, Wahle, and Koletzko 1997, Giuffrida et al. 2016, Jiang et al. 2016, Kovács et al. 2005, Kuipers et al. 2012, López-López et al. 2002, Mihalyi et al. 2015, Mojska et al. 2003, Molto-Puigmarti et al. 2011, Much et al. 2013, Peng et al. 2007, Pugo-Gunsam et al. 2007, Sala-Vila et al. 2005, Serra et al. 1997, Thakkar et al. 2019, Tinoco et al. 2008, Urwin et al. 2013, Wu et al. 2019). Our previous study of a large birth cohort is the only one where AA and DHA values increased from the 6th week to the 6th month (Szabo et al. 2010).

The biggest change in AA and DHA values occur from colostrum to mature human milk with a significant decrease, while the change from colostrum to transitional and from transitional to mature milk is less clear (Table 3). Some studies found a significant decrease of AA from colostrum to transitional milk (Boersma et al. 1991, Genzel-Boroviczény, Wahle, and Koletzko 1997, Giuffrida et al. 2016, Jiang et al. 2016, Kovács et al. 2005, López-López et al. 2002, Molto-Puigmarti et al. 2011, Sala-Vila et al. 2005, Sanchez-Hernandez et al. 2019, Urwin et al. 2013, Wu et al. 2019) and from transitional to mature milk (Boersma et al. 1991, Genzel-Boroviczény, Wahle, and Koletzko 1997, Giuffrida et al. 2016, Kovács et al. 2005, López-López et al. 2002, Molto-Puigmarti et al. 2011, Wu et al. 2019), while others found only a significant difference between colostrum and mature milk (Kuipers et al. 2012, Serra et al. 1997, Thakkar et al. 2019). DHA values decreased also significantly from colostrum to transitional milk (Genzel-Boroviczény, Wahle, and Koletzko 1997, Giuffrida et al. 2016, Jiang et al. 2016, López-López et al. 2002, Molto-Puigmarti et al. 2011, Urwin et al. 2013, Wu et al. 2019) and from transitional to mature milk (Boersma et al. 1991, Genzel-Boroviczény, Wahle, and Koletzko 1997, Giuffrida et al. 2016, Jiang et al. 2016, López-López et al. 2002, Sala-Vila et al. 2005, Serra et al. 1997), but in a few studies only values between colostrum and mature human milk differed (Kovács et al. 2005, Kuipers et al. 2012, Thakkar et al. 2019).

Table 3. Change in arachidonic acid (AA) and docosahexaenoic acid (DHA) values during lactation in some selected articles

First author, Year of publication	Time of sampling	AA	DHA
Antonakou, 2013	1, 3, 6. month (M)	↓ (6. month)	↔
Aydin, 2014	3. day (C); 7. day (T); 28. day (M)	↔	↓
Barreiro, 2018	1, 2, 3, 4, 5. month (M)	↔	↔
Barrera, 2018	1, 2, 3, 4, 5, 6. month (M)	↔	↓ (all time points)
Berenhauser, 2012	2-4. day (C); >16. day (M)	↓	↓
Bobinski, 2013	4-7. day (T); 2. month (M)	↓	↔
Boersma, 1991	0-4. day (C); 5-9. day (T); 10-30. day (M)	↓ (all time points)	↓ (CM, TM)
Decsi, 2000	5. day (C); 1-14. month (M)	↓	↓
Deng, 2018	0-7. day (C); >21. day (M)	↔	↓ (North Jiangsu)
da Costa, 2016	1-3. day (C); 3. month (M)	↓	↓
Genzel-Boroviczeny, 1997	5. day (C); 10. day (T); 20, 30. day (M)	↓ (CT, TM)	↓ (CT, TM)
Giuffrida, 2016	0-5. day (C); 6-15. day (T); 16. day-8. month (M)	↓ (CT, TM)	↓ (CT, TM)
Grote, 2016	1, 2, 3, 6. month (M)	↔	↔
Jiang, 2016	1. day (C); 14. day (T); 42. day (M)	↓ (CT, CM)	↓ (CT, CM, TM)
Khor, 2020	2-3. week (T); 3-8, 8-16. week (M)	↔	↔
Kovacs, 2005	1. day, 4. day (C); 7. day (T); 14, 21. day (M)	↓ (CT, CM, TM)	↓ (CM)
Kuipers, 2012	1-5. day (C); 6-14. day (T); >14. day (M)	↓ (CM)	↓ (CM)
Li, 2009	10-13. day (T); 21-25. day (M)	↔	↓
López-López, 2002	1-6. day (C); 7-15. day (T); >15. day (M)	↓ (CT, CM, TM)	↓ (CT, TM)
Mihalyi, 2015	1. day (C); 6. week, 6. month (M)	↓	↓
Mojska, 2003	3-4. day (C); 5-6, 9-10. week (M)	↓	↓
Molto-Puigmarti, 2011	2-4. day (C); 8-12. day (T); 28-32. day (M)	↓ (CT, CM, TM)	↓ (CT, CM)
Much, 2013	6. week, 16. week (M)	↓	↓
Peng, 2007	5. day (C); 42. day (M)	↓	↓
Pugo-Gunsam, 2007	5. day (C); 42. day (M)	↓	↓
Ribeiro, 2008	7. day (T); 4, 8, 12, 16. week (M)	↓	↔
Rueda, 1998	1-5. day (C); 6-15. day (T); 16-35. day (M)	↔ (Spain) ↓ (Panama)	↔ (Spain, Panama)
Sala-Vila, 2005	1-5. day (C); 6-15. day (T); 15-30. day (M)	↓ (CT, CM)	↓ (CM, TM)
Sanchez-Hernandez, 2019	1-5. day (C); 6-15. day (T); >15. day (M)	↓ (CT, CM)	↔
Serra, 1997	1. day (C); 4, 7. day (T); 14, 21, 28. day (M)	↓ (CM, TM)	↓ (TM)
Szabó, 2010	6. week, 6. month	↑	↑
Thakkar, 2019	(C); (T); (M) -8. week	↓ (CM)	↓ (CM)
Tinoco, 2008	1-5. day (C); 35-42. day (M)	↓	↓
Urwin, 2013	3-5. day (C); 14. day (T); 28. day (M)	↓ (CT, CM)	↓ (CT, CM)
Wu, 2019	1-5. day (C); 11-15. day (T); 41-45. day (M)	↓ (CT, CM, TM)	↓ (CT, CM)

C: colostrum; T: transitional milk; M: mature milk; ↓: significant decrease; ↑: significant increase; ↔ no significant change; CT: significant difference between colostrum and transitional milk; CM: significant difference between colostrum and mature milk; TM: significant difference between transitional milk and mature milk

Since the largest changes in the lipid content of breast milk occur in the first two weeks of breastfeeding, the fatty acid composition may also show a significant change in the first two weeks. Our research group therefore performed a day-to-day study on the fatty acid change during the first week of lactation and found several changes in AA values. These values were constantly decreasing between 1-6th days of lactation, while DHA values remained quite constant (Minda et al. 2004). This was corroborated in another study (Kovács et al. 2005) with a significant decrease of AA between on the one hand 1st and 4th day and on the other hand 7th and 14th day of lactation. In contrast, in this study DHA values became lower from the 1st and 4th day to the 14th day. Similarly, in a Finnish study (Luukkainen, Salo, and Nikkari 1994) both AA and DHA values decreased significantly between the 1st and 12th day of lactation in term as well as in preterm milk samples.

Fat content of milk is increasing in the first two weeks of lactation, so there might be an increase in the absolute concentration of PUFAs also. Most authors publish fatty acid data as weight percent of all determined fatty acids, but some studies publish absolute values also (as mg/dL). Unfortunately, only one study was found (Dai et al. 2020) investigating absolute concentration of human milk samples. In this study a light increase in both AA and DHA concentrations were found, although, no statistical analysis was published. More studies are needed to investigate the change of fatty acids in absolute concentrations during lactation to conclude whether the important n-3 and n-6 metabolites increase, decrease or remain stable in the course of lactation.

Although human milk mainly consists of TGs, other lipid classes are also important in the course of lactation. In a former study (Harzer et al. 1983) lipid classes were investigated in the first 36 days of lactation. Over the investigated time period there was a 1.9-fold increase in TG, 1.7-fold decrease in cholesterol content, while total PL remained quite stable. In spite of the little change in total PL content, there were some marked changes in major PL subclasses: SM and PE increased, while PC decreased. In contrast, Bitman et al. (Bitman et al. 1984) found significant decrease in the PL content in term milk samples during lactation, but the direction of change of main subclasses was similar until the 21st day of lactation (SM increased, PC decreased). Giuffrida et al. (Giuffrida et al. 2016) investigated the different PL classes and gangliosides during lactation and found a significant decrease in PL classes (PC, PE, SM, phosphatidylserine, phoshatidylinositol) over the lactation period, while ganglioside GM3 was significantly increased and GD3 decreased. In Spanish mothers (Sala-Vila et al. 2005) PE and phosphatidylserine fraction significantly increased, while PC significantly decreased. The other investigated major PL fractions (phosphatidylinositol, SM) remained quite constant in the course of lactation. Similarly, in Irish samples the PC significantly decreased, while PE significantly increased from colostrum to mature milk (Ingvordsen Lindahl et al. 2019). Similar to Bitman (Bitman et al. 1984) and Giuffrida (Giuffrida et al. 2016), in this study total PL concentration also significantly decreased during lactation. Wu et al. (Wu et al. 2019) also showed a significant decrease in PC, PE, SM and total PL absolute content from colostrum to mature milk, while relative concentrations decreased significantly in PC and increased in SM fraction.

EFFECT OF PRETERM BIRTH

During the last trimester of pregnancy and in the first months of postnatal life there is an increased lipid accretion due to rapid neurodevelopment. The accumulation of AA and DHA starts prenatally via placental transfer and continues postnatally with the diet. AA content in the forebrain reaches a plateau at about 1-2 years of life, while the accumulation of DHA is continued until 4-6 years in the PE lipids (Martinez and Mougan 1998).

The availability of LCPUFAs is influenced by many factors, like fatty acid stores at birth, the ability of synthesis from their shorter chain precursors as well as dietary sources. The fat stores of premature infants as well as the conversion of LCPUFAs is limited. Although, preterm infants (Salem et al. 1996) as well as very low birth weight preterm infants (Carnielli et al. 1996) are capable to synthesize AA and DHA, this seems to be inadequate. Therefore, this increased demand should be covered by the diet, i.e., breastmilk.

Lot of studies investigated the fatty acid composition of human milk after preterm birth and compared it to term milk samples (**Table 4**). Almost all presented studies investigated the fatty acid composition in colostrum and some of them found significantly higher LA (Al-Tamer and Mahmood 2004, Berenhauser et al. 2012) and AA values (Al-Tamer and Mahmood 2004, Kovács et al. 2005) in preterm milk, while others found no significant differences (Berenhauser et al. 2012, Genzel-Boroviczény, Wahle, and Koletzko 1997, Granot et al. 2016, Kuipers et al. 2011, 2012, Rueda et al. 1998, Thakkar et al. 2019) or even lower AA composition (Jang et al. 2011). The results for DHA are also controversial: some authors found significantly higher values in preterm milk (Kovács et al. 2005, Kuipers et al. 2011, 2012), while others reported no differences (Berenhauser et al. 2012, Genzel-Boroviczény, Wahle, and Koletzko 1997, Granot et al. 2016, Rueda et al. 1998, Thakkar et al. 2019) or ever lower values (Al-Tamer and Mahmood 2004, Jang et al. 2011).

In transitional milk many authors showed no significant differences in the most important PUFA values between the two investigated groups (Genzel-Boroviczény, Wahle, and Koletzko 1997, Rueda et al. 1998), while some found significantly higher DHA (Kovács et al. 2005, Kuipers et al. 2011, 2012) and AA values in preterm samples (Kovács et al. 2005).

In mature milk some studies found no significant differences in AA composition between the two groups (Berenhauser et al. 2012, Genzel-Boroviczény, Wahle, and Koletzko 1997, Thakkar et al. 2019) while others reported significantly higher (Kovács et al. 2005, Kuipers et al. 2011, 2012, Marín et al. 2009) or lower (Rueda et al. 1998) AA values in preterm milk. Luukkainen et al. (Luukkainen, Salo, and Nikkari 1994) compared the fatty acid composition of preterm and term samples in several time points and in mature milk on the 4th week of lactation no differences were found, while on the 12th week DPA and on the 26th week AA, DPA and DHA was increased in preterm samples.

Table 4. Polyunsaturated fatty acid composition in preterm and term human milk samples in some selected studies

First author, year of publ.	Gest. week at birth	Time of sampling	Results
Al-Tamer, 2004	Term: 39.2 ± 3.4 Preterm: 32.7 ± 2.9	Colostrum (3-5 days)	LA, AA, ALA↑; EPA, DHA↓
Berenhauser, 2011	Term: 32.25 ± 2.83 Preterm: 39.7 ± 1.06	Colostrum (2-4 days) Mature milk (16 days-)	C: LA↑; AA, ALA, EPA, DHA↔ M: LA, DHA ↑; AA, ALA, EPA↔
Genzel-Boroviczeny, 1997	Term: 39.91 ± 0.64 Preterm: 28.64 ± 3.72	Colostrum (5 days) Transitional (10 days) Mature (20, 30 days)	C: LA, AA, ALA, EPA, DPA, DHA↔ T: LA, AA, ALA, EPA, DPA, DHA↔ M: LA, AA, ALA, EPA, DPA, DHA↔
Granot, 2016	Term: 39 ± 1.2 Preterm: 31 ± 2.9	Colostrum (4-5 days)	LA, AA, ALA, EPA, DHA↔
Jang, 2011	Term: 39.1 ± 1.1 Preterm: 31.7 ± 3.1	Colostrum (1 week)	LA↔; AA, DHA↓; ALA, EPA, DPA↔
Kovács, 2005	Term: 38.5 ± 2.7 Preterm: 28.0 ± 4.2	Colostrum (1, 4 days) Transitional (7 days) Mature (14, 21 days)	C: LA, ALA, DPA↔; AA, EPA, DHA↑ T: LA, ALA, EPA↔; AA, DPA, DHA↑ M: LA, ALA↔; AA, EPA, DPA, DHA↑
Kuipers, 2011	Term: 37 - 42 w. Preterm: 28 - 36 w.	Colostrum (1-5 days) Transitional (6-14 days) Mature (14- days)	C: AA↔; DHA↑ T: AA↔; DHA↑ M: AA↑; DHA↔
Kuipers, 2012	Term: 39.2 ± 1.4 Preterm: 32.9 ± 2.4	Colostrum (1-5 days) Transitional (6-14 days) Mature (4 week)	C: LA, AA, ALA, EPA, DPA↔; DHA↑ T: LA, AA, ALA, EPA, DPA↔; DHA↑ M: LA, ALA, EPA, DPA, DHA↔; AA↑
Luukkainen, 1994	Term: 40 Preterm: 33	Transitional (1 week): T1 Transitional (2 week): T2 Mature (4, 12, 26 week): M4, M12, M26	T1, 2; M4: LA, AA, ALA, EPA; DPA, DHA↔ M12: LA, AA, ALA, EPA, DHA↔; DPA↑ M26: LA, ALA, EPA↔; AA, DPA, DHA↑
Marín, 2009	Term: 37 - 42 week Preterm: 28 - 36 week	Mature (15 days-3 month)	LA↓; AA, DHA↑; ALA, EPA, DPA↔;
Rueda, 1998	Term: 38.5 ± 1.6 Preterm: 34.0 ± 1.7	Colostrum (1-5 days) Transitional (6-15 days) Mature (16-35 days)	C: LA, AA, EPA, DPA, DHA↔ T: LA, AA, EPA, DPA, DHA↔ M: LA, EPA, DPA, DHA↔; AA↓
Thakkar, 2019	Term: 39.5 ± 1.0 Preterm: 30.8 ± 1.4	Colostrum (≤1 week) Transitional (1-2 week) Mature (2-16 week)	C: LA, AA, EPA, DHA↔; ALA↑ T: LA↑; ALA, AA, EPA, DPA, DHA↔ M: LA↑; AA, ALA, EPA, DPA, DHA ↔

LA: linoleic acid, AA: arachidonic acid, ALA: α-linolenic acid, EPA: eicosapentaenoic acid, DPA: docosapentaenoic acid, DHA: docosahexaenoic acid, C: colostrum: T: transitional milk, M: mature milk, ↔: no significant differences, ↑: fatty acid was significantly higher in preterm milk samples, ↓: fatty acid was significantly lower in preterm milk samples

A systematic review (Bokor, Koletzko, and Decsi 2007) found higher percentage contribution of DHA in preterm than in full-term milk, while data of AA were rather controversial. A recent pooled data analysis (Floris et al. 2020) however failed to corroborate this finding: reviewing data of 55 different studies AA, EPA, DPA and DHA values were comparable in all three stages of lactation (colostrum, transitional and mature milk; Figure 2).

Most authors compare the fatty acid composition of preterm and term milk samples as weight% of total fatty acids, however, the absolute content could also be different. Jang et al. (Jang et al. 2011) found no significant differences between the two groups in LA ALA, EPA and DPA values, while AA and DHA values were significantly lower in preterm samples in weight%. When they determined the absolute values of these fatty acids (mg/dL), in preterm samples significantly higher LA, AA and DPA values were found suggesting a possible concentrating effect to provide more FAs for the increased LCPUFA requirements of preterm babies.

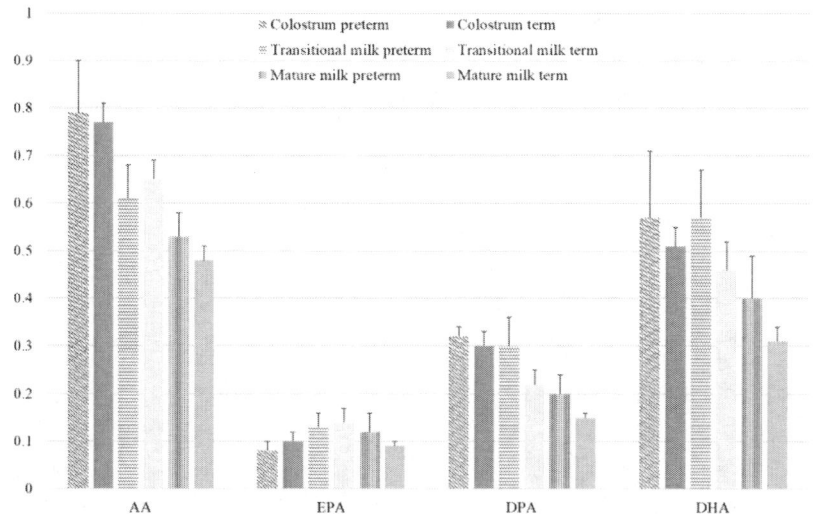

Abbreviations: AA: arachidonic acid, EPA: eicosapentaenoic acid, DPA: docosapentaenoic acid, DHA: docosahexaenoic acid.

Figure 2. LCPUFA composition (g/100 g fatty acid) of preterm and term human milk samples worldwide (data modified according to Floris et al. 2020).

Not only fatty acid composition might differ between preterm and term human milk, but the concentration of different PL classes also. Although there was no significant difference between the PL content in preterm and term human milk samples in all three investigated time points (colostrum, transitional and mature milk), the percentage distribution of sphingomyelin (SM) was higher and that of PC was lower in preterm mature milk compared to term (Shoji et al. 2006). In contrast, in a recent study not only SM concentration was significantly higher in preterm milk in all investigated timepoints compared to term, but PC, PE and total PL concentrations also (Ingvordsen Lindahl et al. 2019).

EFFECT OF MATERNAL DIET

Dietary fatty acids are absorbed from the gut and reach the circulation as TG in the chylomicrons. The absorbed fatty acids could have several metabolic fates: they can be incorporated into cell membranes (as part of PLs), they can be stored in the adipose tissue (as TGs) and they also can be oxidised to gain energy. In addition, they also can serve as educts for the fatty acid elongation and desaturation to synthesize longer chain metabolites (Baker et al. 2016). Therefore, maternal diet can influence the fatty acid composition of breastmilk, but the source of human milk can be different for each fatty acids. According to a tracer study, about 30% of dietary LA was presented in milk (Demmelmair et al. 1998), but about 65% of ALA in mother's milk comes from the diet (Demmelmair et al. 2016).

Diet during Lactation

The fatty acid content of the consumed food, so diet can affect the fatty acid composition of breast milk. DHA is mainly found in sea fishes, so the dietary intake of mothers living far from the sea is lower. This was shown by a review (Fu et al. 2016) reporting significantly higher DHA levels in breastmilk from mothers with accessibility to marine foods than those from mothers without accessibility (0.35 [0.2]% vs. 0.25 [0.14]%; $p < 0.05$; mean [SD]). The strong positive correlation between dietary EPA and DHA on the one hand and n-3 PUFA, EPA and DHA values in breast milk on the other hand raises the role of their dietary intake in the concentration of these fatty acids. In contrast, human milk AA values were not correlated to maternal diet (Kim et al. 2017, Urwin et al. 2013). This strong correlation between dietary and milk DHA values is corroborated by other studies also (Liu et al. 2016, Liu, Liu, and Wang 2019, Aumeistere et al. 2019, Urwin et al. 2013).

In a Chinese study three different parts of China was investigated according to geographical location and different dietary habits. In the northern region (Inner Mongolia) mothers' diet is rich in dairy and ruminant products, in the southeast part (Guangxi) many fresh fruits and vegetables are consumed and in the eastern coastal part (North Jiangsu) diet is very rich in sea products like fishes and shrimps (Deng et al. 2018). Not only the fat intake of the three regions differed significantly, but the food items consumed also varied. The highest DHA level was found at the coastal area (North Jiangsu), and lowest in Inner Mongolia. The fatty acid composition of breastmilk was also influenced by diet: the fat intake was inversely correlated to milk LA, cereal intake negatively to LA and positively to ALA in milk, while increased dairy products and fishes decreased milk monounsaturated fatty acid content. Another Chinese study also found differences in human milk fatty acid content according to geographical location (Giuffrida et al. 2016). Not only the relative concentrations, but the absolute (g/L) values were also different in the human milk of mothers living at different parts of China (Liu, Liu, and Wang 2019). The DHA content of breastmilk was the lowest in mothers living in the inland area (Hohhot) and the highest in the coastal part (Shandong and Guangzhou) not only in colostrum, but in mature milk samples also. Similarly, mothers with high dietary fish intake had higher DHA values in the colostrum and transitional milk, but no differences were found in mature human milk (Jiang et al. 2016). Japanese diet is also rich in marine products and as a result both EPA and DHA content of milk samples were significantly higher than in Nigerian mothers with limited access to DHA (Ogunleye et al. 1991). High dietary intake of sea fishes occurred higher DHA values in human milk compared to mothers with low DHA intake in Chinese mothers too (Peng et al. 2009).

Lactating women in Latvia had significant positive correlations between n-6 PUFA, LA, n-3 PUFA, ALA and DHA intake and the corresponding fatty acid values measured in their milk samples (Aumeistere et al. 2019). In Chilean lactating mothers low n-3 LCPUFA intake during pregnancy and lactation occurred a reduction of DHA levels in breast milk (Barrera et al. 2018).

In Slovenia no differences were found either in the dietary intakes or the fatty acid composition of human milk between lactating women living in urban and rural areas (Fidler, Salobir, and Stibilj 2001). In contrast, another study by the same research group investigated breastfeeding mothers living in three geographically different parts of Slovenia and found some differences in the fatty acid levels in human milk (Fidler, Salobir, and Stibilj 2000). An interesting result was the lack of higher DHA levels in human milk of mothers living in the coastal region compared to the inner regions. The authors explained this by the general availability of sea food throughout the country in the form of either fresh or tinned fish.

In Burkina Faso the fatty acid composition of human milk samples was compared during a period of food shortage (lean season) and during a favourable food availability period (post-harvest season) in both rural and urban area (Thiombiano-Coulibaly et al. 2003). In the rural area significantly higher LA and AA values were found in the post-harvest season, while in the urban area LA values were significantly higher in the lean season. DHA values were higher during the lean season in the rural area while in the urban area higher DHA values were found in the post-harvest season. These differences can be explained, at least in part, by the maternal diet.

Vegan or vegetarian diet can also influence the fatty acid composition of breast milk. As n-3 LCPUFAs (EPA and DHA) are mainly found in fish, therefore breastmilk of vegan lactating women is deficient in DHA

(Karcz and Krolak-Olejnik 2020). In contrast, in Mauritian mothers, who were predominantly vegetarians with low animal fat intake significantly higher LA, AA, ALA, EPA, DPA and DHA values were found than in French lactating women (Pugo-Gunsam et al. 2007).

In a recently published article we reported significant changes in breast milk fatty acid composition over a decade in the same population in two different cohorts (Siziba, Lorenz, et al. 2020). According to the single principal component analysis two principal components were found. The earlier cohort was associated with the first principal component (high saturated, and trans fatty acid (TFA), low n-3 and n-6 LCPUFA content), while the later cohort was associated with the second principal component (high saturated fatty acid, n-3 and n-6 LCPUFA, and low TFA content). Although, no nutritional surveys were conducted in these cohorts, the results suggest the potential influencing effect of dietary habits on the fatty acid composition of human milk.

Potential Effect of Nationality

The effect of nationality can be caused by several factors and only one of them is the different diet. Some other circumstances that can affect breastmilk fatty acid composition are nutritional status, different socio-economical background and different genetics of mothers. Several authors investigated this question and compared the milk fatty acid composition in mothers living in different countries or even different continents or in mothers who live in the same country but have different ethnicity or race.

It has been suggested that mothers in different countries may have different fatty acid compositions in maternal biological samples, such as plasma, red blood cell membranes or human milk, due to different dietary habits, different food preparation methods and different foods available in shops, as well as possible differences in its nutrient content. However, the differences between the milk samples from different countries can be objectively explained only by detailed dietary questionnaires. In mature milk of Egyptian and American mothers different fatty acid composition was found (Borschel et al. 1986). Although no dietary questionnaires were used, there were some obvious differences in their diet. Some n-3 and n-6 fatty acids differed in Spanish and Panamanian mothers also, especially in transitional and mature human milk (Rueda et al. 1998). Mothers living in Panama had significantly higher ALA, DPA and n-3 LCPUFA values, but similar DHA compared to Spanish mothers that might reflect the effect of different diet. In Bolivian mothers, who present a forager-horticulturalist population with minimal access to market foods significantly lower milk LA and higher AA, ALA, EPA, DPA and DHA values were found compared to milk samples of mothers living in the USA (Martin et al. 2012).

Differences in fatty acid composition can be explained much more clearly if different nations have different dietary intakes based on dietary questionnaires. In a study comparing milk samples of Australian, New Zealand and Singaporean women with different dietary habits, several differences were found. Singaporean women eat more fish rich in n-3 fatty acids and this was resulted in significantly higher DHA values in their breastmilk samples compared to Australian and New Zealand mothers (Gao et al. 2018).

When comparing the fatty acid composition of breast milk samples from European and African mothers, data were quite similar despite the different methodologies and dietary intake (Koletzko, Thiel, and Abiodun 1992). The AA values changed between 0.19 (UK) – 1.2 (Poland) in European and between 0.31 (Gambia) – 1.0 (South Africa) in African studies. DHA values also were quite similar: 0.1 (Hungary) – 1.59 (UK) vs. 0.1 (South Africa) – 0.93 (Nigeria). The values of the two essential FAs were also comparable: 6.9 (UK) – 16.35 (Spain) vs. 5.7 (Ivory Coast) – 17.2 (Egypt) for LA and 0.81 (Germany) – <1.8 (Germany) vs. 0.1 (South Africa) – 1.41 (Nigeria) for ALA. But these data should be compared critically because of the different fatty acid determinations: before 1988 with packed and after 1988 with capillary column. The former resolved fewer fatty acids while the latter is more accurate and precise and more fatty acids can be determined.

In a recent systematic review human milk fatty acid content was compared according to continents (Bahreynian, Feizi, and Kelishadi 2020). The highest pooled mean of saturated fatty acids was found in Africa, the highest monounsaturated fatty acids in North America and the highest PUFAs in Asia (Figure 3). The highest TFA content of human milk was also found in North American samples. In the background of these findings could the effect of different dietary habits between population, lifestyles, socioeconomic status of mothers, different genetics as well as several confounding factors, such as different sample collection methods, differences in exact sampling times, duration of exclusive breastfeeding, different estimations of maternal diet, large heterogeneity between studies, maternal age, sample size, year of study and different methods for measuring milk fatty acids. But despite these confounding factors, it is clear that there are some differences in the fatty acid composition of breast milk samples of different nationalities.

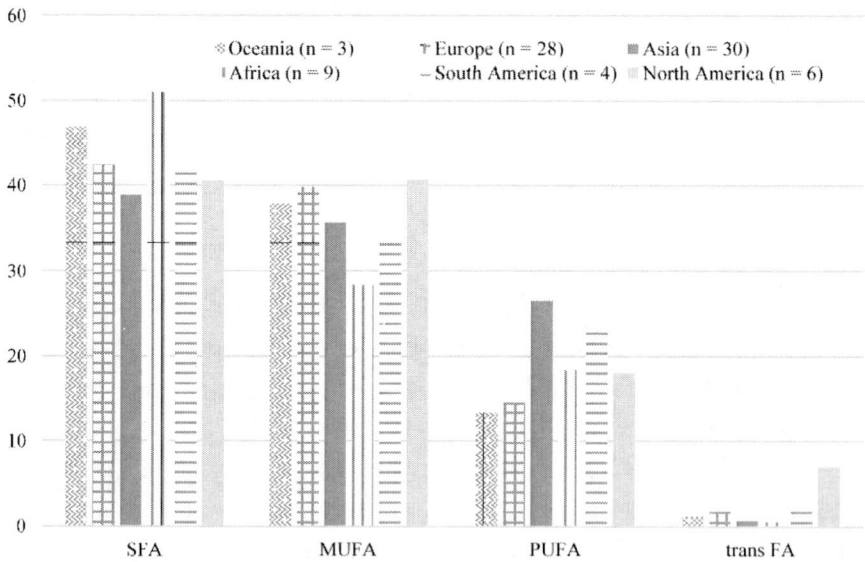

Abbreviations: SFA: saturated fatty acids, MUFA: monounsaturated fatty acid, PUFA: polyunsaturated fatty acids.

Figure 3. Fatty acid content of human milk according to continents (n: number of included studies) (data converted from Bahreynian et al., 2020).

Fu et al. (Fu et al. 2016) also found worldwide region-specific alterations in DHA and AA content of human milk. DHA levels were the highest in Asian, and the lowest in the North American samples. There were some (not significant) differences in the AA content also with relatively high levels in the Eastern Mediterranean region and lower levels in the Oceania region (data mainly from Australia).

When we investigate the human milk fatty acid composition in different nationalities living in the same country we can better study the potential effect of nationality without the disturbing effect of the available food, differences of its nutrient content and other national dietary differences among countries. Although, different nationalities living in the same country might have also different dietary patterns. Our research group investigated human milk fatty acid composition in mothers with different nationality but living in Germany (Szabó et al. 2007). Turkish mothers had the highest LA and lowest DHA levels in their milk samples at the 6th week of lactation and German mothers born in Germany the lowest. Another study also investigated the milk fatty acid composition in different ethnicities (Asian, Maori/Pacific Island and New-Zealand European) with known differences in their dietary intake (Butts et al. 2018). Although there were no significant differences in the macronutrient, water and most mineral content among these ethnicities, several differences were found in the fatty acid content. Asian mothers had the highest PUFA and lowest saturated fatty acid intake and this was resulted in the highest PUFA, DHA, LA, AA content of their breast milk. In colostrum also some differences were found among the three nationalities living in Greece (Greek, Albanian, other), but mainly in saturated and monounsaturated fatty acids (Sinanoglou et al. 2017).

There could be also big differences in the socio-economic background of investigated nationalities resulting in striking differences in the fat content and fatty acid composition of human milk (Rocquelin et al. 2003). In that study mothers with lower education and lower socio-economic status had much higher LA and AA, and much lower ALA and DHA values in their milk samples. This was resulted in a very increased LA/ALA ratio (52.6 ± 27.6).

CONCLUSION

Exclusive breastfeeding is recommended in the first 6 months of life because of its optimal nutrient content. Among others, fatty acids provided by human milk play an important role in the early postnatal neurodevelopment. On the basis of the available literature, the fatty acid composition of human milk can be influenced by many factors, like maternal diet, nationality, gestational age at birth and stage of lactation. Maternal diet mainly influences the n-3 fatty acid content and therefore increased DHA intake during lactation can cause higher milk DHA values. There are also several differences in the fatty acid composition of breastmilk of mothers living on different continents with more preferable ratios in Asian women. However, the effect of preterm birth is less obvious with some differences between human milk of preterm compared to term infants in individual studies but when we combine them and systematically review their results these differences disappear.

REFERENCES

AAP, American Academy of Pediatrics, Section on, Breastfeeding. (2012). "Breastfeeding and the use of human milk." *Pediatrics*, *129* (3), e827-41. doi: 10.1542/peds.2011-3552.

Al-Tamer, Y. Y. & Mahmood, A. A. (2004). "Fatty-acid composition of the colostrum and serum of fullterm and preterm delivering Iraqi mothers." *Eur J Clin Nutr*, *58* (8), 1119-24. doi: 10.1038/sj.ejcn.1601939.

Alashmali, S. M., Hopperton, K. E. & Bazinet, R. P. (2016). "Lowering dietary n-6 polyunsaturated fatty acids: interaction with brain arachidonic and docosahexaenoic acids." *Curr Opin Lipidol*, *27* (1), 54-66. doi: 10.1097/MOL.0000000000000255.

Antonakou, A., Skenderi, K. P., Chiou, A., Anastasiou, C. A., Bakoula, C. & Matalas, A. L. (2013). "Breast milk fat concentration and fatty acid pattern during the first six months in exclusively breastfeeding Greek women." *Eur J Nutr*, *52* (3), 963-73. doi: 10.1007/s00394-012-0403-8.

Aumeistere, L., Ciprovica, I., Zavadska, D., Andersons, J., Volkovs, V. & Celmalniece, K. (2019). "Impact of Maternal Diet on Human Milk Composition Among Lactating Women in Latvia." *Medicina (Kaunas)*, *55* (5). doi: 10.3390/medicina55050173.

Aydin, I., Turan, Ö., Aydin, F. N., Koc, E., Hirfanoglu, İ. M., Akyol, M., Öztosun, M., Akgül, E. Ö., Demirin, H., Kilic, S., Erbil, M. K. & Özgürtas, T. (2014). "Comparing the fatty acid levels of preterm and term breast milk in Turkish women." *Turkish Journal of Medical Sciences*, *44*, 305-310. doi: 10.3906/sag-1302-131.

Bahreynian, M., Feizi, A. & Kelishadi, R. (2020). "Is fatty acid composition of breast milk different in various populations? A systematic review and meta-analysis." *Int J Food Sci Nutr*, *71* (8), 909-920. doi: 10.1080/09637486.2020.1746958.

Baker, E. J., Miles, E. A., Burdge, G. C., Yaqoob, P. & Calder, P. C. (2016). "Metabolism and functional effects of plant-derived omega-3 fatty acids in humans." *Prog Lipid Res*, *64*, 30-56. doi: 10.1016/j.plipres.2016.07.002.

Barreiro, R., Diaz-Bao, M., Cepeda, A., Regal, P. & Fente, C. A. (2018). "Fatty acid composition of breast milk in Galicia (NW Spain): A cross-country comparison." *Prostaglandins Leukot Essent Fatty Acids*, *135*, 102-114. doi: 10.1016/j.plefa.2018.06.002.

Barrera, C., Valenzuela, R., Chamorro, R., Bascunan, K., Sandoval, J., Sabag, N., Valenzuela, F., Valencia, M. P., Puigrredon, C. & Valenzuela, A. (2018). "The Impact of Maternal Diet during Pregnancy and Lactation on the Fatty Acid Composition of Erythrocytes and Breast Milk of Chilean Women." *Nutrients*, *10* (7). doi: 10.3390/nu10070839.

Berenhauser, A. C., Pinheiro do Prado, A. C., da Silva, R. C., Gioielli, L. A. & Block, J. M. (2012). "Fatty acid composition in preterm and term breast milk." *Int J Food Sci Nutr*, *63* (3), 318-25. doi: 10.3109/09637486.2011.627843.

Bergmann, R. L., Haschke-Becher, E., Klassen-Wigger, P., Bergmann, K. E., Richter, R., Dudenhausen, J. W., Grathwohl, D. & Haschke, F. (2008). "Supplementation with 200 mg/day docosahexaenoic acid from mid-pregnancy through lactation improves the docosahexaenoic acid status of mothers with a habitually low fish intake and of their infants." *Ann Nutr Metab*, *52* (2), 157-66. doi: 10.1159/000129651.

Bitman, J., Wood, D. L., Mehta, N. R., Hamosh, P. & Hamosh, M. (1984). "Comparison of the phospholipid composition of breast milk from mothers of term and preterm infants during lactation." *Am J Clin Nutr*, *40* (5), 1103-19. doi: 10.1093/ajcn/40.5.1103.

Bitman, J., Wood, L., Hamosh, M., Hamosh, P. & Mehta, N. R. (1983). "Comparison of the lipid composition of breast milk from mothers of term and preterm infants." *Am J Clin Nutr*, *38* (2), 300-12. doi: 10.1093/ajcn/38.2.300.

Bobinski, R., Mikulska, M., Mojska, H. & Simon, M. (2013). "Comparison of the fatty acid composition of transitional and mature milk of mothers who delivered healthy full-term babies, preterm babies and full-term small for gestational age infants." *Eur J Clin Nutr*, *67* (9), 966-71. doi: 10.1038/ejcn.2013.96.

Boersma, E. R., Offringa, P. J., Muskiet, F. A., Chase, W. M. & Simmons, I. J. (1991). "Vitamin E, lipid fractions, and fatty acid composition of colostrum, transitional milk, and mature milk: an international comparative study". *Am J Clin Nutr*, *53* (5), 1197-204. doi: 10.1093/ajcn/53.5.1197.

Bokor, S., Koletzko, B. & Decsi, T. (2007). "Systematic review of fatty acid composition of human milk from mothers of preterm compared to full-term infants." *Ann Nutr Metab*, *51* (6), 550-6. doi: 10.1159/000114209.

Borschel, M. W., Elkin, R. G., Kirksey, A., Story, J. A., Galal, O., Harrison, G. G. & Jerome, N. W. (1986). "Fatty acid composition of mature human milk of Egyptian and American women." *Am J Clin Nutr*, *44* (3), 330-5. doi: 10.1093/ajcn/44.3.330.

Brenna, J. T., Salem, N., Jr. Sinclair, A. J. & Cunnane, S. C. (2009). Acids International Society for the Study of Fatty, and Issfal Lipids. "alpha-

Linolenic acid supplementation and conversion to n-3 long-chain polyunsaturated fatty acids in humans." *Prostaglandins Leukot Essent Fatty Acids*, *80* (2-3), 85-91. doi: 10.1016/j.plefa.2009.01.004.

Budowski, P., Druckmann, H., Kaplan, B. & Merlob, P. (1994). "Mature milk from Israeli mothers is rich in polyunsaturated fatty acids." *World Rev Nutr Diet*, *75*, 105-8. doi: 10.1159/000423560.

Butts, C. A., Hedderley, D. I., Herath, T. D., Paturi, G., Glyn-Jones, S., Wiens, F., Stahl, B., & Gopal, P. (2018). "Human Milk Composition and Dietary Intakes of Breastfeeding Women of Different Ethnicity from the Manawatu-Wanganui Region of New Zealand." *Nutrients*, *10* (9). doi: 10.3390/nu10091231.

Bzikowska-Jura, A., Sobieraj, P., Szostak-Wegierek, D. & Wesolowska, A. (2020). "Impact of Infant and Maternal Factors on Energy and Macronutrient Composition of Human Milk." *Nutrients*, *12* (9). doi: 10.3390/nu12092591.

Carnielli, V. P., Wattimena, D. J., Luijendijk, I. H., Boerlage, A., Degenhart, H. J. & Sauer, P. J. (1996). "The very low birth weight premature infant is capable of synthesizing arachidonic and docosahexaenoic acids from linoleic and linolenic acids." *Pediatr Res*, *40* (1), 169-74. doi: 10.1203/00006450-199607000-00029.

Clandinin, M. T., Chappell, J. E., Leong, S., Heim, T., Swyer, P. R. & Chance, G. W. (1980a). "Extrauterine fatty acid accretion in infant brain: implications for fatty acid requirements". *Early Hum Dev*, *4* (2), 131-8. doi: 10.1016/0378-3782(80)90016-x.

Clandinin, M. T., Chappell, J. E., Leong, S., Heim, T., Swyer, P. R. & Chance, G. W. (1980b). "Intrauterine fatty acid accretion rates in human brain: implications for fatty acid requirements". *Early Hum Dev*, *4* (2), 121-9. doi: 10.1016/0378-3782(80)90015-8.

Dai, X., Yuan, T., Zhang, X., Zhou, Q., Bi, H., Yu, R., Wei, W. & Wang, X. (2020). "Short-chain fatty acid (SCFA) and medium-chain fatty acid (MCFA) concentrations in human milk consumed by infants born at different gestational ages and the variations in concentration during lactation stages." *Food Funct*, *11* (2), 1869-1880. doi: 10.1039/c9fo02595b.

de Fluiter, K. S., Kerkhof, G. F., Ialp van Beijsterveldt, Breij, L. M., van de Heijning, B. J. M., Abrahamse-Berkeveld, M. & Hokken-Koelega, A. C. S. (2020). "Longitudinal human milk macronutrients, body composition and infant appetite during early life." *Clin Nutr*. doi: 10.1016/j.clnu.2020.11.024.

de Souza Santos da Costa, R., da Silva Santos, F., de Barros Mucci, D., de Souza, T. V., de Carvalho Sardinha, F. L., Moutinho de Miranda Chaves, C. R. & das Gracas Tavares do Carmo, M. (2016). "trans Fatty Acids in Colostrum, Mature Milk and Diet of Lactating Adolescents." *Lipids*, *51* (12), 1363-1373. doi: 10.1007/s11745-016-4206-1.

Decsi, T. (2000). "Fatty acid composition of human milk in Hungary." *Acta Pediatr*, *89*, 1394-5.

Del Prado, M., Villalpando, S., Elizondo, A., Rodríguez, M., Demmelmair, H. & Koletzko, B. (2001). "Contribution of dietary and newly formed arachidonic acid to human milk lipids in women eating a low-fat diet" *Am J Clin Nutr*, *74* (2), 242-7. doi: 10.1093/ajcn/74.2.242.

Delgado-Noguera, M. F., Calvache, J. A., Bonfill Cosp, X., Kotanidou, E. P. & Galli-Tsinopoulou, A. (2015). "Supplementation with long chain polyunsaturated fatty acids (LCPUFA) to breastfeeding mothers for improving child growth and development." *Cochrane Database Syst Rev*, (7), CD007901. doi: 10.1002/14651858.CD007901.pub3.

Demmelmair, H., Baumheuer, M., Koletzko, B., Dokoupil, K. & Kratl, G. (1998). "Metabolism of U 13 C-labeled linoleic acid in lactating women." *J Lipid Res*, *39* (7), 1389-96.

Demmelmair, H. & Koletzko, B. (2018). "Lipids in human milk." *Best Pract Res Clin Endocrinol Metab*, *32* (1), 57-68. doi: 10.1016/j.beem.2017.11.002.

Demmelmair, H., Kuhn, A., Dokoupil, K., Hegele, V., Sauerwald, T. & Koletzko, B. (2016). "Human lactation: oxidation and maternal transfer of dietary (13)C-labelled alpha-linolenic acid into human milk." *Isotopes Environ Health Stud*, *52* (3), 270-80. doi: 10.1080/10256016.2015.1071362.

Deng, L., Zou, Q., Liu, B., Ye, W., Zhuo, C., Chen, L., Deng, Z. Y., Fan, Y. W. & Li, J. (2018). "Fatty acid positional distribution in colostrum and mature milk of women living in Inner Mongolia, North Jiangsu and Guangxi of China." *Food Funct*, 9 (8), 4234-4245. doi: 10.1039/c8fo00787j.

EFSA Panel on Dietetic Products, Nutrition, and Allergies. (2010). "Scientific Opinion on Dietary Reference Values for fats, including saturated fatty acids, polyunsaturated fatty acids, monounsaturated fatty acids, trans fatty acids, and cholesterol." *EFSA Journal*, 8 (3). doi: 10.2903/j.efsa.2010.1461.

ESPGHAN, Committee on Nutrition, Agostoni, C., Braegger, C., Decsi, T., Kolacek, S., Koletzko, B., Michaelsen, K. F., Mihatsch, W., Moreno, L., Puntis, J., Shamir, R., Szajewska, H., Turck, D. & van Goudoever, J. (2009). "Breast-feeding: A commentary by the ESPGHAN Committee on Nutrition" *J Pediatr Gastroenterol Nutr*, 49 (1), 112-25. doi: 10.1097/MPG.0b013e31819f1e05.

Farquharson, J., Cockburn, F., Patrick, W. A., Jamieson, E. C. & Logan, R. W. (1992). "Infant cerebral cortex phospholipid fatty-acid composition and diet." *The Lancet*, 340 (8823), 810-3. doi: 10.1016/0140-6736(92)92684-8.

Fidler, N., Salobir, K. & Stibilj, V. (2000). "Fatty acid composition of human milk in different regions of Slovenia." *Ann Nutr Metab*, 44 (5-6), 187-93. doi: 10.1159/000046682.

Fidler, N., Salobir, K. & Stibilj, V. (2001). "Fatty acid composition of human colostrum in Slovenian women living in urban and rural areas." *Biol Neonate*, 79 (1), 15-20. doi: 10.1159/000047060.

Fischer Fumeaux, C. J., Garcia-Rodenas, C. L., De Castro, C. A., Courtet-Compondu, M. C., Thakkar, S. K., Beauport, L., Tolsa, J. F. & Affolter, M. (2019). "Longitudinal Analysis of Macronutrient Composition in Preterm and Term Human Milk: A Prospective Cohort Study." *Nutrients*, 11 (7). doi: 10.3390/nu11071525.

Floris, L. M., Stahl, B., Abrahamse-Berkeveld, M. & Teller, I. C. (2020). "Human milk fatty acid profile across lactational stages after term and preterm delivery: A pooled data analysis." *Prostaglandins Leukot Essent Fatty Acids*, 156, 102023. doi: 10.1016/j.plefa.2019.102023.

Francois, C. A., Connor, S. L., Bolewicz, L. C. & Connor, W. E. (2003). "Supplementing lactating women with flaxseed oil does not increase docosahexaenoic acid in their milk." *Am J Clin Nutr*, *77* (1), 226-33. doi: 10.1093/ajcn/77.1.226.

Fu, Y., Liu, X., Zhou, B., Jiang, A. C. & Chai, L. (2016). "An updated review of worldwide levels of docosahexaenoic and arachidonic acid in human breast milk by region." *Public Health Nutr*, *19* (15), 2675-87. doi: 10.1017/S1368980016000707.

Gao, C., Gibson, R. A., McPhee, A. J., Zhou, S. J., Collins, C. T., Makrides, M., Miller, J. & Liu, G. (2018). "Comparison of breast milk fatty acid composition from mothers of premature infants of three countries using novel dried milk spot technology." *Prostaglandins Leukot Essent Fatty Acids*, *139*, 3-8. doi: 10.1016/j.plefa.2018.08.003.

Gawlik, N. R., Anderson, A. J., Makrides, M., Kettler, L. & Gould, J. F. (2020). "The Influence of DHA on Language Development: A Review of Randomized Controlled Trials of DHA Supplementation in Pregnancy, the Neonatal Period, and Infancy." *Nutrients*, *12* (10). doi: 10.3390/nu12103106.

Genzel-Boroviczény, O., Wahle, J. & Koletzko, B. (1997). "Fatty acid composition of human milk during the 1st month after term and preterm delivery." *Eur J Pediatr*, *156* (2), 142-7. doi: 10.1007/s004310050573.

Gidrewicz, D. A. & Fenton, T. R. (2014). "A systematic review and meta-analysis of the nutrient content of preterm and term breast milk." *BMC Pediatr*, *14*, 216. doi: 10.1186/1471-2431-14-216.

Giuffrida, F., Cruz-Hernandez, C., Bertschy, E., Fontannaz, P., Masserey Elmelegy, I., Tavazzi, I., Marmet, C., Sanchez-Bridge, B., Thakkar, S. K., De Castro, C. A., Vynes-Pares, G., Zhang, Y. & Wang, P. (2016). "Temporal Changes of Human Breast Milk Lipids of Chinese Mothers." *Nutrients*, *8* (11). doi: 10.3390/nu8110715.

Granot, E., Ishay-Gigi, K., Malaach, L. & Flidel-Rimon, O. (2016). "Is there a difference in breast milk fatty acid composition of mothers of preterm and term infants?" *J Matern Fetal Neonatal Med*, *29* (5), 832-5. doi: 10.3109/14767058.2015.1020785.

Grote, V., Verduci, E., Scaglioni, S., Vecchi, F., Contarini, G., Giovannini, M., Koletzko, B., Agostoni, C. & Project European Childhood Obesity. (2016). "Breast milk composition and infant nutrient intakes during the first 12 months of life." *Eur J Clin Nutr*, *70* (2), 250-6. doi: 10.1038/ejcn.2015.162.

Harzer, G., Haug, M., Dieterich, I. & Gentner, P. R. (1983). "Changing patterns of human milk lipids in the course of the lactation and during the day." *Am J Clin Nutr*, *37* (4), 612-21. doi: 10.1093/ajcn/37.4.612.

Hoffman, D. R., Boettcher, J. A. & Diersen-Schade, D. A. (2009). "Toward optimizing vision and cognition in term infants by dietary docosahexaenoic and arachidonic acid supplementation: a review of randomized controlled trials." *Prostaglandins Leukot Essent Fatty Acids*, *81* (2-3), 151-8. doi: 10.1016/j.plefa.2009.05.003.

Hosseini, M., Valizadeh, E., Hosseini, N., Khatibshahidi, S. & Raeisi, S. (2020). "The Role of Infant Sex on Human Milk Composition." *Breastfeed Med*, *15* (5), 341-346. doi: 10.1089/bfm.2019.0205.

Ingvordsen Lindahl, I. E., Artegoitia, V. M., Downey, E., O'Mahony, J. A., O'Shea, C. A., Ryan, C. A., Kelly, A. L., Bertram, H. C. & Sundekilde, U. K. (2019). "Quantification of Human Milk Phospholipids: the Effect of Gestational and Lactational Age on Phospholipid Composition." *Nutrients*, *11* (2). doi: 10.3390/nu11020222.

Italianer, M. F., Naninck, E. F. G., Roelants, J. A., van der Horst, G. T. J., Reiss, I. K. M., Goudoever, J. B. V., Joosten, K. F. M., Chaves, I. & Vermeulen, M. J. (2020). "Circadian Variation in Human Milk Composition, a Systematic Review." *Nutrients*, *12* (8). doi: 10.3390/nu12082328.

Jackson, D. A., Imong, S. M., Silprasert, A., Ruckphaopunt, S., Woolridge, M. W., Baum, J. D. & Amatayakul, K. (1988). "Circadian variation in fat concentration of breast-milk in a rural northern Thai population." *Br J Nutr*, *59* (3), 349-63. doi: 10.1079/bjn19880044.

Jang, S. H., Lee, B. S., Park, J. H., Chung, E. J., Um, Y. S., Lee-Kim, Y. C. & Kim, E. A. (2011). "Serial changes of fatty acids in preterm breast milk of Korean women." *J Hum Lact*, 27 (3), 279-85. doi: 10.1177/0890334411405059.

Jiang, J., Wu, K., Yu, Z., Ren, Y., Zhao, Y., Jiang, Y., Xu, X., Li, W., Jin, Y., Yuan, J. & Li, D. (2016). "Changes in fatty acid composition of human milk over lactation stages and relationship with dietary intake in Chinese women." *Food Funct*, 7 (7), 3154-62. doi: 10.1039/c6fo00304d.

Karatas, Z., Durmus Aydogdu, S., Dinleyici, E. C., Colak, O. & Dogruel, N. (2011). "Breastmilk ghrelin, leptin, and fat levels changing foremilk to hindmilk: is that important for self-control of feeding?" *Eur J Pediatr*, 170 (10), 1273-80. doi: 10.1007/s00431-011-1438-1.

Karcz, K. & Krolak-Olejnik, B. (2020). "Vegan or vegetarian diet and breast milk composition - a systematic review." *Crit Rev Food Sci Nutr*, 1-18. doi: 10.1080/10408398.2020.1753650.

Khor, G. L., Tan, S. S., Stoutjesdijk, E., Ng, K. W. T., Khouw, I., Bragt, M., Schaafsma, A., Dijck-Brouwer, D. A. J. & Muskiet, F. A. J. (2020). "Temporal Changes in Breast Milk Fatty Acids Contents: A Case Study of Malay Breastfeeding Women." *Nutrients*, 13 (1). doi: 10.3390/nu13010101.

Kim, H., Kang, S., Jung, B. M., Yi, H., Jung, J. A. & Chang, N. (2017). "Breast milk fatty acid composition and fatty acid intake of lactating mothers in South Korea." *Br J Nutr*, 117 (4), 556-561. doi: 10.1017/S0007114517000253.

Koletzko, B. (2016). "Human Milk Lipids." *Ann Nutr Metab*, 69, Suppl 2, 28-40. doi: 10.1159/000452819.

Koletzko, B., Boey, C. C., Campoy, C., Carlson, S. E., Chang, N., Guillermo-Tuazon, M. A., Joshi, S., Prell, C., Quak, S. H., Sjarif, D. R., Su, Y., Supapannachart, S., Yamashiro, Y. & Osendarp, S. J. (2014). "Current information and Asian perspectives on long-chain polyunsaturated fatty acids in pregnancy, lactation, and infancy: systematic review and practice recommendations from an early nutrition academy workshop." *Ann Nutr Metab*, 65 (1), 49-80. doi: 10.1159/000365767.

Koletzko, B., Lien, E., Agostoni, C., Bohles, H., Campoy, C., Cetin, I., Decsi, T., Dudenhausen, J. W., Dupont, C., Forsyth, S., Hoesli, I., Holzgreve, W., Lapillonne, A., Putet, G., Secher, N. J., Symonds, M., Szajewska, H., Willatts, P., Uauy, R. & Group World Association of Perinatal Medicine Dietary Guidelines Working. (2008). "The roles of long-chain polyunsaturated fatty acids in pregnancy, lactation and infancy: review of current knowledge and consensus recommendations." *J Perinat Med*, *36* (1), 5-14. doi: 10.1515/JPM.2008.001.

Koletzko, B., Thiel, I. & Abiodun, P. O. (1992). "The fatty acid composition of human milk in Europe and Africa." *J Pediatr*, *120* (4 Pt 2), S62-70.

Kovács, A., Funke, S., Marosvölgyi, T., Burus, I. & Decsi, T. (2005). "Fatty acids in early human milk after preterm and full-term delivery." *J Pediatr Gastroenterol Nutr*, *41* (4), 454-9. doi: 10.1097/01.mpg.0000176181.66390.54.

Kuipers, R. S., Luxwolda, M. F., Dijck-Brouwer, D. A. & Muskiet, F. A. (2011). "Differences in preterm and term milk fatty acid compositions may be caused by the different hormonal milieu of early parturition." *Prostaglandins Leukot Essent Fatty Acids*, *85* (6), 369-79. doi: 10.1016/j.plefa.2011.08.001.

Kuipers, R. S., Luxwolda, M. F., Dijck-Brouwer, D. A. & Muskiet, F. A. (2012). "Fatty acid compositions of preterm and term colostrum, transitional and mature milks in a sub-Saharan population with high fish intakes." *Prostaglandins Leukot Essent Fatty Acids*, *86* (4-5), 201-7. doi: 10.1016/j.plefa.2012.02.006.

Lacombe, R. J. S., Chouinard-Watkins, R. & Bazinet, R. P. (2018). "Brain docosahexaenoic acid uptake and metabolism." *Mol Aspects Med*, *64*, 109-134. doi: 10.1016/j.mam.2017.12.004.

Lassek, W. D. & Gaulin, S. J. (2014). "Linoleic and docosahexaenoic acids in human milk have opposite relationships with cognitive test performance in a sample of 28 countries." *Prostaglandins Leukot Essent Fatty Acids*, *91* (5), 195-201. doi: 10.1016/j.plefa.2014.07.017.

Lessen, R. & Kavanagh, K. (2015). "Position of the academy of nutrition and dietetics: promoting and supporting breastfeeding." *J Acad Nutr Diet*, *115* (3), 444-9. doi: 10.1016/j.jand.2014.12.014.

Li, J., Fan, Y., Zhang, Z., Yu, H., An, Y., Kramer, J. K. & Deng, Z. (2009). "Evaluating the trans fatty acid, CLA, PUFA and erucic acid diversity in human milk from five regions in China." *Lipids*, *44* (3), 257-71. doi: 10.1007/s11745-009-3282-x.

Liu, M. J., Li, H. T., Yu, L. X., Xu, G. S., Ge, H., Wang, L. L., Zhang, Y. L., Zhou, Y. B., Li, Y., Bai, M. X. & Liu, J. M. (2016). "A Correlation Study of DHA Dietary Intake and Plasma, Erythrocyte and Breast Milk DHA Concentrations in Lactating Women from Coastland, Lakeland, and Inland Areas of China." *Nutrients*, *8* (5), doi: 10.3390/nu8050312.

Liu, Y., Liu, X. & Wang, L. (2019). "The investigation of fatty acid composition of breast milk and its relationship with dietary fatty acid intake in 5 regions of China." *Medicine (Baltimore)*, *98* (24), e15855. doi: 10.1097/MD.0000000000015855.

López-López, A., López-Sabater, M. C., Campoy-Folgoso, C., Rivero-Urgell, M. & Castellote-Bargalló, A. I. (2002). "Fatty acid and sn-2 fatty acid composition in human milk from Granada (Spain) and in infant formulas." *Eur J Clin Nutr*, *56* (12), 1242-54. doi: 10.1038/sj.ejcn.1601470.

Luukkainen, P., Salo, M. K. & Nikkari, T. (1994). "Changes in the fatty acid composition of preterm and term human milk from 1 week to 6 months of lactation." *J Pediatr Gastroenterol Nutr*, *18* (3), 355-60. doi: 10.1097/00005176-199404000-00018.

Makrides, M., Neumann, M. A., Byard, R. W., Simmer, K. & Gibson, R. A. (1994). "Fatty acid composition of brain, retina, and erythrocytes in breast- and formula-fed infants." *Am J Clin Nutr*, *60* (2), 189-94. doi: 10.1093/ajcn/60.2.189.

Makrides, M., Simmer, K., Goggin, M. & Gibson, R. A. (1993). "Erythrocyte docosahexaenoic acid correlates with the visual response of healthy, term infants". *Pediatr Res*, *33* (4 Pt 1), 425-7. doi: 10.1203/00006450-199304000-00021.

Mangel, L., Morag, S., Mandel, D., Marom, R., Moran-Lev, H. & Lubetzky, R. (2020). "The Effect of Infant's Sex on Human Milk Macronutrients Content: An Observational Study." *Breastfeed Med*, *15* (9), 568-571. doi: 10.1089/bfm.2020.0228.

Marín, M. C., Sanjurjo, A. L., Sager, G., Margheritis, C. & de Alaniz, M. J. (2009). "[Fatty acid composition of human milk from mothers of preterm and full-term infants]" *Arch Argent Pediatr*, *107* (4), 315-20. doi: 10.1590/S0325-00752009000400008.

Martin, M. A., Lassek, W. D., Gaulin, S. J., Evans, R. W., Woo, J. G., Geraghty, S. R., Davidson, B. S., Morrow, A. L., Kaplan, H. S. & Gurven, M. D. (2012). "Fatty acid composition in the mature milk of Bolivian forager-horticulturalists: controlled comparisons with a US sample." *Matern Child Nutr*, *8* (3), 404-18. doi: 10.1111/j.1740-8709.2012.00412.x.

Martinez, M. A. & Mougan, I. (1998). "Fatty acid composition of human brain phospholipids during normal development." *J Neurochem*, *71* (6), 2528-33. doi: 10.1046/j.1471-4159.1998.71062528.x.

Mihalyi, K., Gyorei, E., Szabo, E., Marosvolgyi, T., Lohner, S. & Decsi, T. (2015). "Contribution of n-3 long-chain polyunsaturated fatty acids to human milk is still low in Hungarian mothers." *Eur J Pediatr*, *174* (3), 393-8. doi: 10.1007/s00431-014-2411-6.

Minda, H., Kovacs, A., Funke, S., Szasz, M., Burus, I., Molnar, S., Marosvolgyi, T. & Decsi, T. (2004). "Changes of fatty acid composition of human milk during the first month of lactation: a day-to-day approach in the first week." *Ann Nutr Metab*, *48* (3), 202-9. doi: 10.1159/000079821.

Mojska, H., Socha, P., Socha, J., Soplinska, E., Jaroszewska-Balicka, W. & Szponar, L. (2003). "Trans fatty acids in human milk in Poland and their association with breastfeeding mothers' diets." *Acta Paediatr*, *92* (12), 1381-7. doi: 10.1080/08035250310006692.

Molto-Puigmarti, C., Castellote, A. I., Carbonell-Estrany, X. & Lopez-Sabater, M. C. (2011). "Differences in fat content and fatty acid proportions among colostrum, transitional, and mature milk from women delivering very preterm, preterm, and term infants." *Clin Nutr*, *30* (1), 116-23. doi: 10.1016/j.clnu.2010.07.013.

Much, D., Brunner, S., Vollhardt, C., Schmid, D., Sedlmeier, E. M., Bruderl, M., Heimberg, E., Bartke, N., Boehm, G., Bader, B. L., Amann-Gassner, U. & Hauner, H. (2013). "Breast milk fatty acid profile in relation to infant growth and body composition: results from the INFAT study." *Pediatr Res*, *74* (2), 230-7. doi: 10.1038/pr.2013.82.

Ogunleye, A., Fakoya, A. T., Niizeki, S., Tojo, H., Sasajima, I., Kobayashi, M., Tateishi, S. & Yamaguchi, K. (1991). "Fatty acid composition of breast milk from Nigerian and Japanese women." *J Nutr Sci Vitaminol (Tokyo)*, *37* (4), 435-42. doi: 10.3177/jnsv.37.435.

Paulaviciene, I. J., Liubsys, A., Molyte, A., Eidukaite, A. & Usonis, V. (2020). "Circadian changes in the composition of human milk macronutrients depending on pregnancy duration: a cross-sectional study." *Int Breastfeed J*, *15* (1), 49. doi: 10.1186/s13006-020-00291-y.

Peng, Y. M., Zhang, T. Y., Wang, Q., Zetterstrom, R. & Strandvik, B. (2007). "Fatty acid composition in breast milk and serum phospholipids of healthy term Chinese infants during first 6 weeks of life." *Acta Paediatr*, *96* (11), 1640-5. doi: 10.1111/j.1651-2227.2007.00482.x.

Peng, Y., Zhou, T., Wang, Q., Liu, P., Zhang, T., Zetterstrom, R. & Strandvik, B. (2009). "Fatty acid composition of diet, cord blood and breast milk in Chinese mothers with different dietary habits." *Prostaglandins Leukot Essent Fatty Acids*, *81* (5-6), 325-30. doi: 10.1016/j.plefa.2009.07.004.

Perrin, M. T., Belfort, M. B., Hagadorn, J. I., McGrath, J. M., Taylor, S. N., Tosi, L. M. & Brownell, E. A. (2020). "The Nutritional Composition and Energy Content of Donor Human Milk: A Systematic Review." *Adv Nutr*, *11* (4), 960-970. doi: 10.1093/advances/nmaa014.

Pugo-Gunsam, Prity, Philippe Guesnet, Anwar Hussein Subratty, Dev Anand Rajcoomar, Chantal Maurage. & Charles Couet. (2007). "Fatty acid composition of white adipose tissue and breast milk of Mauritian and French mothers and erythrocyte phospholipids of their full-term breast-fed infants." *British Journal of Nutrition*, *82* (4), 263-271. doi: 10.1017/s0007114599001464.

Qi, K., Hall, M. & Deckelbaum, R. J. (2002). "Long-chain polyunsaturated fatty acid accretion in brain" *Curr Opin Clin Nutr Metab Care*, *5* (2), 133-8. doi: 10.1097/00075197-200203000-00003.

Ribeiro, M., Balcao, V., Guimaraes, H., Rocha, G., Moutinho, C., Matos, C., Almeida, C., Casal, S. & Guerra, A. (2008). "Fatty acid profile of human milk of Portuguese lactating women: prospective study from the 1st to the 16th week of lactation." *Ann Nutr Metab*, *53* (1), 50-6. doi: 10.1159/000156597.

Rocquelin, G., Tapsoba, S., Kiffer, J. & Eymard-Duvernay, S. (2003). "Human milk fatty acids and growth of infants in Brazzaville (The Congo) and Ouagadougou (Burkina Faso)." *Public Health Nutr*, *6* (3), 241-8. doi: 10.1079/PHN2002420.

Rueda, R., Ramírez, M., García-Salmerón, J. L., Maldonado, J. & Gil, A. (1998). "Gestational age and origin of human milk influence total lipid and fatty acid contents." *Ann Nutr Metab*, *42* (1), 12-22. doi: 10.1159/000012713.

Sala-Vila, A., Castellote, A. I., Rodriguez-Palmero, M., Campoy, C. & Lopez-Sabater, M. C. (2005). "Lipid composition in human breast milk from Granada (Spain): changes during lactation." *Nutrition*, *21* (4), 467-73. doi: 10.1016/j.nut.2004.08.020.

Salem, N., Jr. Wegher, B., Mena, P. & Uauy, R. (1996). "Arachidonic and docosahexaenoic acids are biosynthesized from their 18-carbon precursors in human infants." *Proc Natl Acad Sci U S A.*, *93* (1), 49-54. doi: 10.1073/pnas.93.1.49.

Sanchez-Hernandez, S., Esteban-Munoz, A., Gimenez-Martinez, R., Aguilar-Cordero, M. J., Miralles-Buraglia, B. & Olalla-Herrera, M. (2019). "A Comparison of Changes in the Fatty Acid Profile of Human Milk of Spanish Lactating Women during the First Month of Lactation Using Gas Chromatography-Mass Spectrometry. A Comparison with Infant Formulas." *Nutrients*, *11* (12). doi: 10.3390/nu11123055.

Serra, G., Marletta, A., Bonacci, W., Campone, F., Bertini, I., Lantieri, P. B., Risso, D. & Ciangherotti, S. (1997). "Fatty acid composition of human milk in Italy." *Biol Neonate*, 72 (1), 1-8. doi: 10.1159/000244459.

Sherry, C. L., Oliver, J. S. & Marriage, B. J. (2015). "Docosahexaenoic acid supplementation in lactating women increases breast milk and plasma docosahexaenoic acid concentrations and alters infant omega 6:3 fatty acid ratio." *Prostaglandins Leukot Essent Fatty Acids*, 95, 63-9. doi: 10.1016/j.plefa.2015.01.005.

Shoji, H., Shimizu, T., Kaneko, N., Shinohara, K., Shiga, S., Saito, M., Oshida, K., Shimizu, T., Takase, M. & Yamashiro, Y. (2006). "Comparison of the phospholipid classes in human milk in Japanese mothers of term and preterm infants." *Acta Paediatr*, 95 (8), 996-1000. doi: 10.1080/08035250600660933.

Simopoulos, Artemis P. (2010). "The omega-6/omega-3 fatty acid ratio: health implications." *Oléagineux, Corps gras, Lipides*, 17 (5), 267-275. doi: 10.1051/ocl.2010.0325.

Sinanoglou, V. J., Cavouras, D., Boutsikou, T., Briana, D. D., Lantzouraki, D. Z., Paliatsiou, S., Volaki, P., Bratakos, S., Malamitsi-Puchner, A. & Zoumpoulakis, P. (2017). "Factors affecting human colostrum fatty acid profile: A case study." *PLoS One*, 12 (4), e0175817. doi: 10.1371/journal.pone.0175817.

Siziba, L. P., Chimhashu, T., Siro, S. S., Ngounda, J. O., Jacobs, A., Malan, L., Smuts, C. M. & Baumgartner, J. (2020). "Breast milk and erythrocyte fatty acid composition of lactating women residing in a periurban South African township." *Prostaglandins Leukot Essent Fatty Acids*, 156, 102027. doi: 10.1016/j.plefa.2019.102027.

Siziba, L. P., Lorenz, L., Brenner, H., Carr, P., Stahl, B., Mank, M., Marosvolgyi, T., Decsi, T., Szabo, E., Rothenbacher, D. & Genuneit, J. (2020). "Changes in human milk fatty acid composition and maternal lifestyle-related factors over a decade: a comparison between the two Ulm Birth Cohort Studies." *Br J Nutr*, 1-8. doi: 10.1017/S0007114520004006.

Szabo, E., Boehm, G., Beermann, C., Weyermann, M., Brenner, H., Rothenbacher, D. & Decsi, T. (2010). "Fatty acid profile comparisons in human milk sampled from the same mothers at the sixth week and the sixth month of lactation." *J Pediatr Gastroenterol Nutr, 50* (3), 316-20. doi: 10.1097/MPG.0b013e3181a9f944.

Szabó, E., Boehm, G., Beermann, C., Weyermann, M., Brenner, H., Rothenbacher, D. & Decsi, T. (2007). "trans Octadecenoic acid and trans octadecadienoic acid are inversely related to long-chain polyunsaturates in human milk: results of a large birth cohort study." *Am J Clin Nutr, 85* (5), 1320-6. doi: 10.1093/ajcn/85.5.1320.

Thakkar, S. K., De Castro, C. A., Beauport, L., Tolsa, J. F., Fischer Fumeaux, C. J., Affolter, M. & Giuffrida, F. (2019). "Temporal Progression of Fatty Acids in Preterm and Term Human Milk of Mothers from Switzerland." *Nutrients, 11* (1). doi: 10.3390/nu11010112.

Thiombiano-Coulibaly, N., Rocquelin, G., Eymard-Duvernay, S., Kiffer-Nunes, J., Tapsoba, S. & Traore, S. A. (2003). "Seasonal and environmental effects on breast milk fatty acids in Burkina Faso and the need to improve the omega 3 PUFA content." *Acta Paediatrica, 92* (12), 1388-1393. doi: 10.1111/j.1651-2227.2003.tb00820.x.

Tinoco, S. M., Sichieri, R., Setta, C. L., Moura, A. S. & Md do Carmo. (2008). "Trans fatty acids from milk of Brazilian mothers of premature infants." *J Paediatr Child Health, 44* (1-2), 50-6. doi: 10.1111/j.1440-1754.2007.01172.x.

Urwin, H. J., Zhang, J., Gao, Y., Wang, C., Li, L., Song, P., Man, Q., Meng, L., Froyland, L., Miles, E. A., Calder, P. C. & Yaqoob, P. (2013). "Immune factors and fatty acid composition in human milk from river/lake, coastal and inland regions of China." *Br J Nutr, 109* (11), 1949-61. doi: 10.1017/S0007114512004084.

WHO, World Health Organization. (2018). *Guideline: counselling of women to improve breastfeeding practices.* https://www.who.int/publications/i/item/9789241550468 (accessed on March 02, 2021)

Wu, K., Gao, R., Tian, F., Mao, Y., Wang, B., Zhou, L., Shen, L., Guan, Y. & Cai, M. (2019). "Fatty acid positional distribution (sn-2 fatty acids) and phospholipid composition in Chinese breast milk from colostrum to mature stage." *Br J Nutr, 121* (1), 65-73. doi: 10.1017/S0007114518002994.

In: Human Milk
Editor: John I. Cole

ISBN: 978-1-53619-713-6
© 2021 Nova Science Publishers, Inc.

Chapter 3

LIFE-COURSE HEALTH IMPACTS OF THE NUTRITIONAL CONTENT OF HUMAN MILK IN DIFFERENT STEPS OF LACTATION

Parisa Iravani[1], Motahar Heidari-Beni[2],*
and Roya Kelishadi[3]

[1]Assistant Professor of Pediatrics, Child Growth and Development Research Center, Research Institute for Primordial Prevention of Non-Communicable Disease, Isfahan University of Medical Sciences, Isfahan, Iran
[2]Assistant Professor of Nutrition, Department of Nutrition, Child Growth and Development Research Center, Research Institute for Primordial Prevention of Non-Communicable Disease, Isfahan University of Medical Sciences, Isfahan, Iran
[3]Professor of Pediatrics, Department of Pediatrics, Child Growth and Development Research Center, Research Institute for Primordial Prevention of Non-Communicable Disease, Isfahan University of Medical Sciences, Isfahan, Iran

---

* Corresponding Author's E-mails: motahar.heidari@ nutr.mui.ac.ir and heidari.motahar@ gmail.com.

ABSTRACT

Breastfeeding has numerous health benefits both for the infants and the nursing mothers. In addition to its short-term benefits, it is documented that breastfeeding has preventative roles against chronic non-communicable diseases including hypertension, obesity, diabetes, hyperlipidemia, and cardiovascular diseases in adulthood. These health benefits of human milk are correlated with its nutritional and bioactive components including antimicrobial substances, anti-inflammatory components and hormones, which modulate the body metabolism and composition.

Breastfeeding duration is also one of the factors that might determine the amount of biological effects of milk. The composition of milk changes constantly during lactation to provide nutritional necessities to protect infant against potential harmful pathogens. In addition, gestational length, maternal diseases and nutrition influence the composition of human milk.

The protein content of human milk gradually reduces from the second to the sixth or seventh month of lactation and stabilizes at the final step. Protein concentration is associated with body weight gain of infant and immune protection, as well as increase in nutrient digestion and availability of micronutrient. The fat content increases during lactation and would affect neurological development and cognitive outcome of infant. Human milk carbohydrates impact on the appetite regulation and body composition, which would protect infant against obesity.

Understanding the human milk composition over time and its health benefits can be important for primordial prevention of non-communicable diseases.

This chapter aims to summarize the current literature on the composition of human milk and its life-course functional effects on health outcomes.

Keywords: human milk; composition; health; nutrients; bioactive components; prevention

INTRODUCTION

There has been a considerable increase in various ranges of chronic non-communicable diseases (NCDs) including heart disease, obesity, type II diabetes, cancer, allergies and other immune diseases, asthma and chronic lung diseases, mental illness, and chronic liver and renal diseases. Many of these disorders are linked to prenatal, antenatal and early nutrition. Several studies assessed the effects of early nutrition and nutritional impact in the first 2 years of life on NCDs [1-4].

The World Health Organization (WHO) advises that infants exclusively be breastfed up to the completion of six months. Breastfeeding can continue with complementary foods until two years old [1]. Human milk is an individual-specific biofluid with different nutritional and bioactive components [2-4]. The composition of human milk has evolved over time for providing a well-balanced nutrition and protection against potential infectious pathogens and development of the neonatal immune system. Some factors including time of lactation, length of gestation, maternal diseases, genotype and diet affect the composition of human milk. The mean protein content is regularly reduced from the second to the sixth-seventh month of lactation, and then it becomes stabilized. The preterm milk content is higher and richer than term milk [2-4]. Protein and n-6 long chain polyunsaturated fatty acids reduced during the first three months of lactation. Insulin-like growth factor (IGF)-II and adiponectin concentrations are correlated with fat and protein content [5].

The various bioactive proteins, growth factors, cells, and other constituents in human milk is vital for development of immune system to protect the term/preterm newborn against pathogenic organisms [6]. Among the immunologically important bioactive factors present in human milk, lactoferrin (lactotransferrin) as important biologically active factors directly and indirectly protect the neonate against infection caused by bacteria and other microorganisms. The concentration of lactoferrin in human milk is dependent on lactation stage; colostrum includes more than 5 g/L, which then highly reduces to 2-3 g/L in mature milk [7].

This chapter highlights the recent evidences on the benefits of human milk and its compositions as well as its life-course functional effects on health outcomes. We discussed about the role of nutritional content of human milk in health and disease with a focus on the current trends, important challenges and future perspectives.

ROLE OF NUTRITIONAL CONTENT OF HUMAN MILK IN HEALTH AND DISEASE

Breastfeeding reduces the incidence and/or severity of various infectious diseases such as bacterial meningitis, bacteremia, diarrhea, respiratory tract infection, necrotizing enterocolitis, otitis media, urinary tract infection, and late-onset sepsis in preterm infants. Additionally, breastfeeding have numerous health benefits for the prevention of non-communicable diseases (NCDs). It may reduce the risk of type II diabetes and overweight/obesity, and also may improve the IQ. Post-neonatal infant mortality rates decrease in breastfed infants [8-11]. In addition to the benefits for the infant, breastfeeding also has benefits for the mother. Breastfeeding can reduce the risks of ovarian/premenopausal breast cancers, type 2 diabetes (via the improvement of glucose hormones), the development of hypertension/cardiovascular diseases, and also can help weight lose [12-14]. The exclusive breastfeeding could provide remarkably lower risk of first-time febrile urinary tract infection (UTI). A longer duration of breastfeeding could offer a lower risk of infection after weaning. The protective role of breastfeeding is strongly after birth, then reduce until seven months of age [10].

Human milk influences on the growth, development and functions of the epithelium, immune system or nervous system of the gastrointestinal tract. It can also affects the growth of intestinal villi, the development of intestinal disaccharides (enzymes for breaking down the complex sugars), the permeability of the gastrointestinal tract and resistance to certain inflammatory/immune-mediated diseases [15].

carbohydrates, protein and fat concentration in human milk can differentially affect the development of infant body composition in the first 12 months postpartum, and may potentially affect the risk of later obesity through modulation of body composition [16].

Furthermore, it was indicated that interleukin (IL)-10 as one of the cytokine in human milk, protected premature infants against necrotizing enterocolitis [15]. Proteins such as lactoferrin, secretory IgA, κ-casein, lactoperoxidase, haptocorrin, lactadherin and peptides that are generated from human milk proteins during digestion can obstruct the growth of pathogenic bacteria and viruses. Additionally, lactoferrin, bile-salt stimulated lipase, haptocorrin, κ-casein, and folate-binding proteins can accelerate the absorption of nutrients in the neonatal gut. Remarkably, the proteins in human milk provide suitable amounts of essential amino acids and can promote proper growth in infant [17].

According to findings of clinical studies, human milk (with suitable fortification for the very low-birth-weight infant) is the standard care for preterm, as well as term infants with short and long-term health benefits [18, 19]. Human milk does not provide enough nutrients for the very low birth weight infant can lead to slow growth with the risk of neurocognitive impairment and other poor health outcomes, including retinopathy and bronchopulmonary dysplasia. Thus, human milk should be supplemented (fortified) with the nutrients including protein, calcium, and phosphate to meet the high requirements of nutrients. Though the fortification of human milk is considerably adopted in the neonatal intensive care unit (NICU), there is still much inconsistency and skepticism about fortified human milk.. Individualization and the quality of the fortifiers are very important and should be investigated in more studies [18, 19].

HUMAN MILK OLIGOSACCHARIDES

Human milk oligosaccharides are a family of structurally various unconjugated glycans, which can be found widely in human milk. These prebiotics can act as metabolic substrates for probiotic bacteria and shape intestinal microbiota composition with health advantages for the breast-fed neonate [20]. These oligosaccharides have antiadhesive antimicrobials effects which serve as soluble decoy receptors. They prevent pathogen attachment to infant mucosal surfaces, decrease the risk of pathogenic infections, adjust epithelial and immune cell responses, decrease excessive mucosal leukocyte infiltration and activation and reduce the risk of necrotizing enterocolitis [20]. Study on human milk oligosaccharides in mothers with very preterm infants (< 32 weeks of gestational age, < 1500 g of birthweight) and term infants showed that in preterm milk the levels of α-1,2-linked fructose was lower and the levels of sialylated oligosaccharides especially 3′-sialyllactose was higher than term milk [21]. The largest differences were detected around 40 weeks of postmenstrual age, when the milk of term infants contained the highest concentrations of human milk oligosaccharides [21].

It was reported that human milk oligosaccharides stimulated semi-maturation of human monocytes-derived dendritic cells and enhanced levels of IL-10, IL-27 and IL-6 and reduced levels of IL-12p70 and tumor necrosis factor alpha (TNF-α) [22].

HUMAN MILK FAT

Human milk fat varies from the milk fat of other mammals and even more from vegetable oils. Human milk has a unique complex natural lipid mixtures with beneficial fatty acid composition, distribution, and numerous complex lipids [23, 24]. The amount of human milk fat is different from 3.5% to 4.5% during lactation. Lipid content increases and phospholipids and cholesterol decrease during lactation [25]. Human milk contains 39% monounsaturated fatty acid, 35% saturated fatty acid, and 18% n-6 long-chain polyunsaturated fatty acids (LCPUFA); palmitic acid (16:0), oleic acid (18:0 n-9), and linoleic acid (18:2 n-6) [26]. LCPUFA, particularly arachidonic acid and docosahexaenoic acid are important in human milk [27]. Human milk arachidonic acid levels are associated with dietary habits and regions. Arachidonic acid is considered as a precursor to eicosanoids and endocannabinoids, which are very important nutrients during infancy and childhood [27].

milk fat can supply crucial nutrients, including LCPUFAs, phospholipids, and cholesterol, which are critical for developing brain and cognitive functions during the first year of life [28]. Notably, the amount of fat in human milk is significantly variable, as it fluctuates during the day. It depends on some maternal parameters such as body mass index, age, diet, parity number, and smoking status [29]. Human milk fat has important effects on brain, digestive system, and the maturity of exocrine pancreatic function during infancy. Thus the complex physico-biochemical procedures of milk fat digestion in infancy should be carefully noted. Secretion of pancreatic triglyceride lipase, phospholipase A2, and bile salts is immature in the first few months of life. Pancreatic lipase-related protein 2 and bile salt-stimulated lipase from milk help fat digestion [30]. The macronutrient and energy contents of human milk decrease during the first 6 months of lactation were evaluated. The fat content in hind milk is higher than foremilk. Hind milk have 25-35 kcal/100 ml energy that is higher than foremilk [31].

According to findings, human milk fat globule-EGF factor 8 protein (MFG-E8) could reduce inflammation and might have a preventive effect against Parkinson's disease [32].

Bioactive proteins from milk fat globule membrane (MFGM) have various important functions in cellular procedures and defense mechanisms in infants. There are some important MFGM proteins, including lactotransferrin, beta-casein, lipoprotein lipase, fatty acid synthase, and butyrophilin subfamily 1 member A1 in human milk [33].

HUMAN MILK MICROBIOTA

Human milk microbiota has short and long term infant health benefits. The milk microbiota is originated from the mother's skin, infant's mouth, and also transfers from the maternal gastrointestinal (GI) tract [34]. Generally, the common bacterial taxa in human milk are Staphylococcus and Streptococcus. However several other genera can be detected such as anaerobic Lactobacillus, Bifidobacterium, and Bacteroides. The milk microbiome is significantly variable and potentially is affected by geographic location, delivery mode, time postpartum, feeding mode, social networks, environment, maternal diet, and milk composition [35]. Milk microbiota composition is associated with several maternal parameters including BMI, parity, type of delivery and breastfeeding practices [36]. One of the important factors is geographical location, which can directly impact on composition of breast milk including microbiota and lipids [37]. Among vaginally delivered women, Spanish women had highest amount of Bacteroidetes, while Chinese women had highest amount of Actinobacteria. Spanish and South African women who had cesarean section had higher amount of Proteobacteria and lipid, amino acid and carbohydrate metabolism. Findings about the association between the lipid profile and the microbiota demonstrated that monounsaturated fatty acids were negatively associated with Proteobacteria [37].

It was reported that human milk peptides could selectively promote the growth of bifidobacteria [38]. Some proteins in human milk have antimicrobial influences and hydrolysis of peptides with the gastrointestinal proteases pepsin, trypsin and chymotrypsin do not lead to the loss of bifidogenic performance. Prebiotic lactoferrin-derived peptide-I, PRELP-I, was identified to promote the growth of bifidobacteria as effectively as the native peptides. [38].

HUMAN MILK PROTEINS

Human milk contains various types of proteins with important functions. Some of them such as lactoferrin, bile salt-stimulated lipase, haptocorrin, amylase, α1-antitrypsin, and β-casein can help digestion and utilization of micronutrients and macronutrients in the milk [17, 38, 39]. Various proteins with antimicrobial performances, including haptocorrin, immunoglobulins, κ-casein, lactoferrin, lysozyme, lactoperoxidase, and α-lactalbumin are moderately impervious against proteolysis in the gastrointestinal tract. They provide suitable defense against pathogenic bacteria/viruses and enough nutritional compounds in breastfed infants. In addition, human milk proteins have prebiotic activity and stimulate the growth of valuable microorganisms including Bifidobacteria and Lactobacilli. They may restrict the growth of various pathogenic organisms by reducing the pH of intestine. Additionally, several proteins in human milk including lactoferrin or cytokines can provide the immunomodulatory advantages. Some of them including epidermal growth factor, lactoferrin, and insulin-like growth factor may influence on developing process of the intestinal mucosa and other organs of newborns [17, 39].

Lactoferrin concentration is dependent on duration of lactation. Finding of a study showed that the mean of lactoferrin concentration was the lowest in 1-12 months of lactation (3.39 ± 1.43 g/L) and remarkably increased in the 13-18 months of lactation. This concentration remains at a comparable level in 19-24 month and over 24 months. Additionally, the concentration of lactoferrin in mother's milk demonstrated a positive correlation with protein concentration over lactation from the first to the 48th month [40].

Insulin-like growth factors (IGF-I and -II) are peptide growth factors, which associated with the growth-promoting features of human milk. IGFs in extracellular fluids are associated with high-affinity binding proteins (IGFBPs). one investigation showed that IGF-I and -II, and the IGFBP in human milk were stable under heat treatment conditions [41]. the levels of vascular endothelial growth factor (VEGF), basic fibroblast growth factor (b-FGF), IGF-I and platelet-derived growth factor (PDGF) in human milk were dramatically high. There are different levels of VEGF, b-FGF, IGF-I and PDGF in milk of mothers with preterm and term neonates [42].

HUMAN MILK HORMONES

Hormones such as leptin, adiponectin and ghrelin in the breast milk, can regulate long-term appetite signaling and may be associated with the protection of children against obesity (Table 1). Any mutation in these hormones or their receptors or alterations in their circadian rhythms lead to disorders in brain pathways and control of satiety. At the final step, these changes are associated with the pathogenesis of obesity and metabolic syndrome [43].

Table 1. Some important hormones in breast milk

Hormones	Functions	Refs.
Leptin	Anorexigenic influences; inhibition of fatty acids and triglycerides formation; increase in the oxidation of fatty acids	[44]
Adiponectin	The improvement of insulin sensitivity and fatty acid oxidation and inhibition of hepatic glucose formation; production of anti-inflammatory and anti-atherogenic influences	[45]
Resistin	Antagonizes insulin action, inhibits adipocyte differentiation	[46]
Ghrelin	The orexigenic influence, stimulates growth hormone secretion, inhibits lipolysis stimulates adipogenesis, modulates insulin secretion and gastrointestinal motility	[47, 48]
Obestatin	Anorexigenic influence	[49]

DIFFERENCES IN MILK CONTENTS DURING LACTATION

The amount of phospholipids including sphingomyelin, phosphatidylserine, phosphatidylethanolamine, phosphatidylinositol, and phosphatidylcholine and fatty acid composition of triacylglycerol and phospholipid fractions in breast milk change during lactation [50]. The levels of sphingomyelin are constant during lactation. Content of phospholipids and phosphatidylcholine in mature milk are lower than in colostrum and transitional milk. Additionally, triacylglycerol in mature human milk had lower percentages of docosahexaenoic acid, arachidonic acid, and nervonic acid in comparison with colostrum and mature milk. Long-chain polyunsaturated fatty acids reduce during lactation proceeded [50]. The differences in milk lactoferrin (multifunctional protein) were evaluated during lactation. The concentration of lactoferrin in milk of women delivering at term was 3.16 g/L, 1.73 g/L and 0.90 g/L for colostrum, transitional milk, and mature milk, respectively. It was observed that the concentrations of this protein altered dramatically between stages of lactation (colostrum vs. transitional milk, colostrum vs. mature milk, transitional milk vs. mature milk) [51]. Additionally, the content of β-casomorphin-5 and -7 in human milk was determined in various phases of lactation. As a result, remarkable concentration of these examined β-casomorphins was detected in colostrum than in mature milk;

the level of these peptides can influence on the procedure of maturation of neonates [52].

The human milk metabolomes at the early (9-24 days after delivery) and late (31-87 days after delivery) stages of lactation were noticeably dissimilar. It was observed that milk bacterial communities were typically complex, and exhibited individual-specific profiles [53, 54].

CONCLUSION

Human milk influences on infant health in the short and long term. Studies showed the importance of human milk in relation to gastrointestinal function, growth, neurological and immune system developments for the infant. The changes in the concentration of various components in human milk can be influenced by genetics, physiological/environmental conditions, and nutritional factors. Bioactive factors in human milk including hormones, growth factors, neuropeptides, and anti-inflammatory and immunomodulatory agents increase infant survival rates, decrease the prevalence of NCDs in childhood and adulthood and can support normal growth and development of infants. Thus, because of these benefits, human milk is the first and most important source of nutrition and protection for the infant.

Early life is an appropriate time for intervention. Prevention is the beneficial approach to reduce the prevalence of NCDs. Thus, improving health status through the promotion of breastfeeding which provide wide social and economic benefits is very crucial.

REFERENCES

[1] WHO/UNICEF. *Global strategy for infant and Young Child feeding*. Geneva: World Health Organization; (2003).

[2] Andreas, N. J. Kampmann, B. K., Mehring Le-Doare, Human breast milk: a review on its composition and bioactivity, *Early Hum. Dev.*, 91 (2015) 629-635.

[3] Gidrewicz, D. A. T. Fenton, R. A systematic review and metaanalysis of the nutrient content of preterm and term breast milk, *BMC Pediatr.*, 14 (2014) 216.

[4] Sharp, J. A., V. Modepalli, A.K. Enjapoori, S. Bisana, H.E. Abud, C. Lefevre, K.R. Nicholas, Bioactive Functions of Milk Proteins: a Comparative Genomics Approach, *Journal of Mammary Gland Biology and Neoplasia volume 19* (2014) 289–302.

[5] Grunewald, M., C. Hellmuth, F. F. Kirchberg, M. L. Mearin, R. Auricchio, G. Castillejo, I. R. Korponay-Szabo, I. Polanco, M. Roca, S. L. Vriezinga, K. Werkstetter, B. Koletzko, H. Demmelmair, *Variation and Interdependencies of Human Milk Macronutrients, Fatty Acids, Adiponectin, Insulin, and IGF-II in the European PreventCD Cohort Nutrients*, 11 (2019) 2034.

[6] Ballard,O., A. L. Morrow, Human milk composition: Nutrients and bioactive factors, *Pediatr. Clin. N. Am.*, 60 (2013) 49–74.

[7] Czosnykowska-Łukacka, M., M. Orczyk-Pawiłowicz, B. Broers, B. Królak-Olejnik, *Lactoferrin in Human Milk of Prolonged Lactation Nutrients*, 11 (2019) 2350.

[8] Heinig, M. J., Host defense benefits of breastfeeding for the infant. Effectof breastfeeding duration and exclusivity, *Pediatr. Clin. North Am.*, 48 (2001) 105–123.

[9] Bhandari, N., R. Bahl, S. Mazumdar, J. Martines, R.E. Black, M.K. Bhan, Effect of community-based promotion of exclusive breastfeeding ondiarrhoeal illness and growth: a cluster randomized controlled trial.Infant Feeding Study Group, *Lancet,* 361 (2003) 1418–1423.

[10] Marild, S., S. Hansson, U. Jodal, A. Oden, K. Svedberg, Protective effect ofbreastfeeding against urinary tract infection, *Acta Paediatr,* 93 (2004) 164–168.

[11] Gdalevich, M., D. Mimouni, M. Mimouni, Breast-feeding and the risk of bronchial asthma in childhood: a systematic review with meta-analysis of prospective studies, *J. Pediatr.*, 139 (2001) 261–266.

[12] Danforth, K. N., S. S. Tworoger, J. L. Hecht, B. A. Rosner, G. A. Colditz, S. E. Hankinson, Breastfeeding and risk of ovarian cancer in two prospective cohorts, *Cancer Causes & Control* 18 (2007) 517–523.

[13] Martin, R. M., N. Middleton, D. Gunnell, C. G. Owen, G. D. Smith, Breast-feeding and cancer: the Boyd Orr cohort and a systematic review with meta-analysis, *J. Natl. Cancer Inst.*, 97 (2005) 1446-1457.

[14] Horta, B., C. Victora, W.H. Organization, *Long-term effects of breastfeeding: a systematic review*, World Health Organization, https://apps.who.int/iris/handle/10665/79198 (2013).

[15] Goldman, A. S., Modulation of the Gastrointestinal Tract of Infants by Human Milk. Interfaces and Interactions. An Evolutionary Perspective, *The Journal of Nutrition*, 130 (2000) 426S–431S.

[16] Gridneva, Z., A. Rea, W. Jun Tie, C. Tat Lai, S. Kuganathan, L. C. Ward, K. Murray, P. E. Hartmann, D. T. Geddes, *Carbohydrates in Human Milk and Body Composition of Term Infants during the First 12 Months of Lactation, Nutrients*, 11 (2019) 1472.

[17] Lönnerdal, B., Human Milk Proteins, in: L.K. Pickering, A.L. Morrow, G.M. Ruiz-Palacios, R.J. Schanler (Eds.) Protecting Infants through Human Milk. *Advances in Experimental Medicine and Biology*, Springer, Boston, MA, 2004, pp. 11-25.

[18] Bertino, E., P. Di Nicola, F. Giuliani, C. Peila, E. Cester, C. Vassia, A. Pirra, P. Tonetto, A. Coscia, Benefits of human milk in preterm infant feeding, *J. Pediatr. Neonat. Individual Med.*, 1 (2012) 19-24.

[19] Cerasani, J., F. Ceroni, V. De Cosmi, A. Mazzocchi, D. Morniroli, P. Roggero, F. Mosca, C. Agostoni, M. L. Giannì, *Human Milk Feeding and Preterm Infants' Growthand Body Composition: A Literature Review, Nutrients*, 12 (2020) 1155.

[20] Bode, L., Human milk oligosaccharides: Every baby needs a sugar mama, *Glycobiology*, 22 (2012) 1147–1162.

[21] Austin, S., C. A. De Castro, N. Sprenger, A. Binia, M. Affolter, C. L. Garcia-Rodenas, L. Beauport, J.-F. Tolsa, C. J. Fischer Fumeaux, *Human Milk Oligosaccharides in the Milk of Mothers Delivering Term versus Preterm Infants Nutrients*, 11 (2019) 1282.

[22] Xiao, L., W. R P H van De Worp, R. Stassen, C. van Maastrigt, N. Kettelarij, B. Stahl, B. Blijenberg, S. A. Overbeek, G. Folkerts, J. Garssen, B. van't Land, Human milk oligosaccharides promote immune tolerance via direct interactions with human dendritic cells, *European Journal of Immunology*, 49 (2019) 1001-1014.

[23] Wei, W., Q. Jin, X. Wang, Human milk fat substitutes: Past achievements and current trends, *Progress in Lipid Research*, 74 (2019) 69-86.

[24] Demmelmair, H., B. Koletzko, *Lipids in human milk, Best Practice & Research Clinical Endocrinology & Metabolism*, 32 (2018) 57-68.

[25] Pietrzak-Fiecko, R., A. M. Kamelska-Sadowska, *The Comparison of Nutritional Value of Human Milkwith Other Mammals' Milk, Nutrients*, 12 (2020) 1404.

[26] Floris, L. M., B. Stahl, M. Abrahamse-Berkeveld, I. C. Teller, Human milk fatty acid profile across lactational stages after term and preterm delivery: a pooled data analysis, *Prostaglandins Leukot Essent Fatty Acids*, 156 (2020) 102023.

[27] Salem, N., P. Van Dael, *Arachidonic Acid in Human Milk, Nutrients*, 12 (2020) 626.

[28] Jiao, J., Q. Li, J. Chu, W. Zeng, M. Yang, S. Zhu, Effect of n-3 PUFA supplementation on cognitive function throughout the life span from infancy to old age: a systematic review and meta-analysis of randomized controlled trials, *Am. J. Clin. Nutr.*, 100 (2014) 1422-1436.

[29] Bachour, P., R. Yafawi, F. Jaber, E. Choueiri, Z. Abdel-Razzak, Effects of smoking, mother's age, body mass index, and parity number on lipid, protein, and secretory immunoglobulin A concentrations of human milk, *Breastfeed Med.*, 7 (2012) 179-188.

[30] He, X., S. McClorry, O. Hernell, B. Lönnerdal, C.M. Slupsky, Digestion of human milk fat in healthy infants, *Nutrition Research*, 83 (2020) 15-29.

[31] Saarela, T., J. Kokkonen, M. Koivisto, Macronutrient and energy contents of human milk fractions during the first six months of lactation, *Acta Paediatrica*, 94 (2005) 1176-1181.

[32] Nakashima Y., C. Miyagi-Shiohira, H. Noguchi, T. Omasa, The Healing Effect of Human Milk Fat Globule-EGF Factor 8 Protein (MFG-E8) in *A Rat Model of Parkinson's Disease Brain Sci.*, 8 (2018) 167.

[33] Juvarajah, T., W. I. Wan-Ibrahim, A. Ashrafzadeh, S. Othman, O. Haji Hashim, S. Yee Fung, P. S. Abdul-Rahman, Human Milk Fat Globule Membrane Contains Hundreds of Abundantly Expressed and Nutritionally Beneficial Proteins That Are Generally Lacking in Caprine Milk, *Breastfeeding Medicine*, 13 (2018) https://doi.org/10.1089/bfm.2018.0057.

[34] Fitzstevens, J. L., K. C. Smith, J. I. Hagadorn, M. J. Caimano, A. P. Matson, E. A. Brownell, Systematic Review of the Human Milk Microbiota, *Nutrition in Clinical Practice*, 32 (2017) 354-364.

[35] Demmelmair, H., E. Jiménez, M. C. Collado, S. Salminen, M. K. McGuire, Maternal and Perinatal Factors Associated with the Human Milk Microbiome, *Current Developments in Nutrition*, 4 (2020) nzaa027.

[36] Moossavi, S., S. Sepehri, B. Robertson, L. Bode, S. Goruk, C. J. Field, L. M. Lix, R. J. de Souza, A. B. Becker, P. J. Mandhane, S. E. Turvey, P. Subbarao, T. J. Moraes, D. L. Lefebvre, M. R. Sears, E. Khafipour, M. B. Azad, Composition and Variation of the Human Milk Microbiota Are Influenced by Maternal and Early-Life Factors, *Cell Host & Microbe*, 25 (2019) 324-335.e324.

[37] Kumar, H., E. du Toit, A. Kulkarni, J. Aakko, K. M. Linderborg, Y. Zhang, M. P. Nicol, E. Isolauri, B. Yang, M. C. Collado, S. Salminen, Distinct Patterns in Human Milk Microbiota and Fatty Acid Profiles Across Specific Geographic Locations, *Front. Microbiol.*, https://doi.org/10.3389/fmicb.2016.01619 (2016).

[38] Liepke, C., Adermann K, M. Raida, H. J. Mägert, W.G. Forssmann, H. D. Zucht, Human milk provides peptides highly stimulating the growth of bifidobacteria, *European Journal of Biochemistry*, 269 (2002) 712-718.

[39] Lönnerdal, B., Nutritional and physiologic significance of human milk proteins, *The American Journal of Clinical Nutrition*, 77 (2003) 1537S–1543S.

[40] Lönnerdal, B., Bioactive Proteins in Human Milk: Mechanisms of Action, *The Journal of Pediatrics*, 156 (2010) S26-S30.

[41] Donovan, S. M., R. L. Hintz, R. G. Rosenfeld, Insulin-like growth factors I and II and their binding proteins in human milk: effect of heat treatment on IGF and IGF binding protein stability, *Journal of Pediatric Gastroenterology and Nutrition*, 13 (1991) 242-253.

[42] Ozgurtas, T., I. Aydin, O. Turan, E. Koc, I. Hirfanoglu, M, C. H. Acikel, M. Akyol, M. K. Erbil, Vascular endothelial growth factor, basic fibroblast growth factor, insulin-like growth factor-I and platelet-derived growth factor levels in human milk of mothers with term and preterm neonates, *Cytokine*, 50 (2010) 192-194.

[43] Savino, F., S. Benetti, S. A. Liguori, M. Sorrenti, L. Cordero di montezemolo, Advances on human milk hormones and protection against obesity, *Cell. Mol. Biol.*, 59 (2013) 89-98.

[44] Casabiell, X., V. Pineiro, M. A. Tome, R. Peino, C. Dieguez, F. F. Casanueva, Presence of leptin in colostrum and/or breast milk from lactating mothers: a potential role in the regulation of neonatal food intake, *J. Clin. Endocrinol. Metab.*, 82 (1997) 4270–4273.

[45] Bronsky, J., M. Karpísek, E. Bronská, M. Pechová, B. Jancíková, H. Kotolová, D. Stejskal, R. Prusa, J. Nevoral, Adiponectin, adipocyte fatty acid binding protein, and epidermal fatty acid binding protein: proteins newly identified in human breast milk, *Clin. Chem.*, 52 (2006) 1763-1770.

[46] Ilcol, Y. O., Z. B. Hizli, E. Eroz, Resistin is present in human breast milk and it correlates with maternal hormonal status and serum level of C-reactive protein, *Clin. Chem. Lab. Med.*, 46 (2008) 118-124.

[47] Armstrong, J. J., J. Reilly, Breastfeeding and lowering the risk of childhood obesity, *Lancet* 359 (2002) 2003-2004.

[48] Aydin, S., S. Aydin, Y. Ozkan, S. Kumru, Ghrelin is present in human colostrum, transitional and mature milk, *Peptides,* 27 (2006) 878-882.

[49] Aydin, S., Y. Ozkan, F. Erman, B. Gurates, N. Kilic, R. Colak, T. Gundogan, Z. Catak, M. Bozkurt, O. Akin, Y. Sen, I. Sahn, Presence of obestatin in breast milk: relationship among obestatin, ghrelin, and leptin in lactating women, *Nutrition,* 24 (2008) 689-693.

[50] Sala-Vila, A., A. I. Castellote, M. Rodriguez-Palmero, C. Campoy, M. C. López-Sabater, Lipid composition in human breast milk from Granada (Spain): Changes during lactation, *Nutrition,* 21 (2005) 467-473.

[51] Yang, Z., R. Jiang, Q. Chen, J. Wang, Y. Duan, X. Pang, S. Jiang, Y. Bi, H. Zhang, B. Lönnerdal, J. Lai, S. Yin, *Concentration of Lactoferrin in Human Milk and Its Variation during Lactation in Different Chinese Populations Nutrients,* 10 (2018) 1235.

[52] Jarmołowska, B., K. Sidor, M. Iwan, K. Bielikowicz, M. Kaczmarski, E. Kostyra, H. Kostyra, Changes of β-casomorphin content in human milk during lactation, *Peptides,* 28 (2007) 1982-1986.

[53] Wu, J., M. Domellöf, A. M. Zivkovic, G. Larsson, A. Öhman, M. L. Nording, NMR-based metabolite profiling of human milk: A pilot study of methods for investigating compositional changes during lactation, *Biochemical and Biophysical Research Communications,* 469 (2016) 626-632.

[54] Martín, M.-J., S. Martín-Sosa, L.-A. García-Pardo, P. Hueso, Distribution of Bovine Milk Sialoglycoconjugates During Lactation, *Journal of Dairy Science,* 84 (2001) 995-1000.

INDEX

A

acid, vii, viii, 2, 5, 12, 15, 22, 31, 33, 37, 46, 60, 61, 69, 70, 72, 74, 75, 77, 79, 80, 81, 82, 83, 84, 85, 86, 87, 88, 89, 90, 91, 92, 93, 94, 95, 96, 97, 98, 99, 100, 101, 102, 103, 104, 105, 106, 107, 114, 115, 118, 122, 124
acquired immunity, 2, 20
adhesion, 10, 14, 17, 44
adipocyte, 118, 124
adiponectin, 24, 111, 117
adipose tissue, 85, 104
adulthood, ix, 110, 119
age, 2, 33, 70, 76, 90, 105, 112, 114, 122
alimentary canal, 14, 18, 23
amino acid, 12, 14, 18, 19, 37, 39, 49, 113, 116
amino acids, 12, 14, 18, 19, 37, 39, 113
amniotic fluid, 67
antibiotic, 22, 28, 31, 55
antimicrobial factors, 2, 26
arachidonic acid, viii, 69, 70, 72, 79, 83, 84, 96, 98, 99, 114, 118, 122

B

bacteremia, 21, 112
bacteria, vii, viii, 2, 3, 5, 10, 17, 18, 19, 20, 21, 22, 23, 25, 27, 28, 30, 31, 33, 36, 37, 39, 42, 43, 44, 45, 46, 48, 49, 54, 56, 60, 62, 63, 65, 66, 67, 111, 113, 116
bacterial infection, 39
bacterial strains, 31, 59
bacteriocins, 6, 36, 37, 39, 50, 55, 60, 65
benefits, vii, ix, 1, 4, 24, 46, 47, 110, 111, 112, 113, 115, 119, 120
bile, 23, 40, 48, 50, 113, 115, 116
bioactive components, ix, 3, 48, 49, 110, 111
bioavailability, 47
biological activity, 47, 52, 58
biological fluids, 2
biological samples, 88
blood, 7, 28, 42, 76, 88, 104
blood circulation, 28
body composition, ix, 96, 104, 110, 112
body mass index, 114, 122
body weight, ix, 110

brain, 73, 75, 76, 92, 95, 102, 105, 115, 117
breast cancer, 112
breast feeding, vii, 1, 11
breast milk, viii, 2, 3, 4, 9, 10, 12, 13, 18, 20, 21, 31, 34, 45, 48, 49, 51, 56, 57, 59, 60, 61, 62, 64, 65, 66, 67, 70, 75, 80, 85, 87, 88, 89, 90, 91, 93, 94, 98, 100, 102, 104, 105, 106, 107, 116, 117, 118, 120, 124, 125
breastfeeding, viii, ix, 4, 6, 17, 27, 29, 31, 45, 56, 63, 65, 67, 69, 70, 76, 80, 87, 90, 92, 93, 96, 102, 103, 107, 110, 112, 115, 119, 120, 121
brevis, 41, 49, 60, 67
bronchial asthma, 121
bronchopulmonary dysplasia, 113

C

cancer, 40, 110, 121
cancer cells, 40
cancerous cells, 50
carbohydrate, 7, 17, 37, 42, 72, 116
carbohydrate metabolism, 116
carbohydrates, ix, 3, 15, 110, 112
cardiovascular disease, ix, 110, 112
casein, 10, 12, 14, 16, 17, 21, 112, 115, 116
cell membranes, 40, 72, 85, 88
cell surface, 10, 44
cerebral cortex, 75, 76, 97
childhood, 65, 114, 119, 121, 125
children, 34, 60, 66, 117
Chinese women, 100, 116
cholesterol, 23, 44, 67, 71, 80, 97, 114
circadian rhythms, 117
circulation, 12, 20, 85
cognitive function, 114, 122
cognitive performance, 74
colostrum, vii, ix, 1, 2, 5, 6, 7, 13, 14, 15, 16, 19, 24, 30, 32, 38, 49, 51, 52, 53, 56, 59, 70, 77, 78, 79, 81, 82, 83, 84, 85, 86, 91, 92, 94, 96, 97, 101, 103, 106, 107, 111, 118, 124, 125
composition, vii, viii, ix, x, 3, 5, 6, 8, 13, 15, 21, 25, 27, 29, 31, 32, 36, 45, 48, 52, 53, 55, 57, 58, 60, 61, 62, 63, 70, 75, 77, 80, 82, 83, 84, 85, 86, 87, 88, 89, 90, 91, 92, 93, 94, 95, 96, 97, 98, 99, 100, 101, 102, 103, 104, 105, 106, 107, 110, 111, 112, 113, 114, 115, 118, 120, 121, 123, 125
compounds, 3, 10, 34, 36, 44, 45, 46, 116
consumption, 7, 34, 54
culture, 5, 25, 26, 28, 29, 36, 40, 56
culture media, 25, 26
cytokines, 4, 10, 11, 14, 21, 23, 33, 116
cytomegalovirus, 20
cytotoxicity, 42, 64

D

data analysis, 77, 84, 97, 122
data processing, 26
dendritic cell, 28, 30, 114, 122
developing brain, 114
diabetes, ix, 110, 112
diarrhea, 18, 49, 66, 112
diet, vii, viii, 2, 34, 45, 58, 69, 70, 71, 73, 81, 82, 85, 86, 87, 88, 89, 90, 92, 95, 96, 97, 100, 102, 104, 111, 114, 115
dietary fat, 102
dietary habits, 32, 86, 88, 89, 90, 104, 114
dietary intake, 85, 87, 89, 91, 100
digestion, ix, 13, 14, 18, 23, 29, 39, 110, 113, 115, 116
digestive enzymes, 40
diseases, ix, 4, 21, 29, 36, 45, 110, 111, 112
distribution, 47, 85, 96, 107, 114
diversity, 25, 26, 29, 32, 35, 44, 51, 55, 56, 71, 102
docosahexaenoic acid, viii, 69, 70, 72, 79, 83, 84, 92, 94, 95, 98, 101, 102, 105, 106, 114, 118

E

eicosapentaenoic acid, 72, 83, 84
electrophoresis, 26, 27, 51
energy, 6, 7, 14, 70, 72, 77, 85, 115, 123
environment, 2, 6, 14, 40, 115
environmental conditions, 119
environmental effects, 107
enzymes, 4, 9, 14, 40, 42, 72, 75, 112
essential fatty acids, 72, 73, 74, 76
exopolysaccharides, 42, 43, 44, 58, 63, 64
external environment, 42

F

fat, ix, 7, 12, 15, 16, 23, 58, 70, 71, 72, 77, 81, 86, 88, 91, 92, 96, 99, 100, 103, 110, 111, 112, 114, 115, 122, 123
fat content, ix, 12, 70, 71, 77, 80, 91, 103, 110, 115
fat intake, 86, 88
fatty acid, v, vii, viii, 10, 23, 34, 37, 61, 62, 69, 70, 72, 73, 74, 75, 76, 77, 80, 81, 82, 83, 84, 85, 86, 87, 88, 89, 90, 91, 92, 93, 94, 95, 96, 97, 98, 99, 100, 101, 102, 103, 104, 105, 106, 107, 111, 114, 115, 116, 118, 120, 122, 123, 124
fibroblast growth factor, 117, 124
fish, 61, 86, 87, 89, 94, 101
food, vii, 1, 22, 37, 38, 40, 44, 48, 59, 62, 66, 72, 85, 86, 87, 88, 91, 124
food intake, 124
food production, 38
food products, 62, 66
formation, 19, 39, 40, 44, 118
formula, 4, 18, 29, 32, 38, 46, 47, 57, 58, 63, 76, 102
functional food, 57, 68
fungi, 37, 38, 40

G

gastrointestinal tract, 9, 19, 21, 31, 47, 112, 116
geographical location, 32, 53, 86, 115
gestation, 33, 66, 111
gestational age, vii, viii, 25, 32, 57, 70, 92, 94, 95, 105, 113
gland, 6, 12, 13, 20, 24, 27
glucose, 7, 14, 42, 46, 112, 118
growth, 3, 6, 7, 15, 16, 17, 18, 21, 22, 24, 26, 35, 38, 40, 44, 52, 57, 61, 65, 75, 77, 96, 104, 105, 111, 112, 113, 116, 117, 118, 119, 120, 124
growth factor, 3, 6, 7, 15, 16, 24, 111, 116, 117, 119, 124

H

health, v, vii, viii, ix, x, 2, 3, 6, 26, 32, 34, 35, 44, 45, 49, 54, 56, 62, 65, 69, 96, 98, 105, 106, 107, 109, 110, 111, 112, 113, 115, 119, 121
health effects, 34
health status, 3, 119
heart disease, 110
hormones, ix, 24, 110, 112, 117, 118, 119, 124
human, vii, viii, ix, x, 1, 2, 4, 5, 6, 8, 10, 11, 12, 13, 15, 16, 17, 18, 19, 20, 22, 23, 24, 25, 26, 27, 28, 30, 31, 32, 35, 38, 41, 45, 46, 47, 49, 50, 52, 53, 54, 56, 57, 58, 59, 60, 64, 65, 66, 67, 69, 70, 71, 72, 73, 74, 77, 78, 80, 82, 83, 84, 85, 86, 87, 88, 90, 91, 92, 94, 95, 96, 97, 98, 99, 100, 101, 102, 103, 104, 105, 106, 107, 110, 111, 112, 113, 114, 115, 116, 117, 118, 119, 121, 122, 123, 124, 125
human body, 72
human brain, viii, 69, 95, 103
human immunodeficiency virus, 33

human milk, v, vii, viii, ix, x, 1, 2, 5, 8, 9, 11, 12, 13, 15, 16, 17, 18, 19, 20, 22, 23, 24, 25, 26, 27, 28, 30, 31, 32, 35, 38, 45, 46, 47, 49, 50, 51, 52, 53, 54, 55, 56, 57, 58, 59, 61, 62, 63, 64, 65, 66, 67, 69, 70, 71, 72, 73, 74, 77, 78, 80, 82, 83, 84, 85, 86, 87, 88, 90, 91, 92, 93, 94, 95, 96, 97, 98, 99, 100, 101, 102, 103, 104, 105, 106, 107, 109, 110, 111, 112, 113, 114, 115, 116, 117, 118, 119, 120, 121, 122, 123, 124, 125

human pathogens, vii, 2

humoral immunity, 53

hydrogen, 6, 23, 36, 37, 38

hydrogen peroxide, 6, 23, 36, 37, 38

I

immune modulation, 40

immune response, 21

immune system, 3, 6, 15, 20, 23, 62, 111, 112, 119

immunity, 3, 6, 8, 11, 20, 22, 29, 54, 59

immunoglobulins, 2, 8, 14, 20, 21, 65, 116

immunological memory, 8

immunomodulation, 36

immunomodulatory, 24, 32, 34, 51, 116, 119

immunomodulatory agent, 119

in vitro, 17, 18, 22, 35, 39, 49, 50, 61

infancy, 53, 100, 101, 114, 115, 122

infant mortality, 112

infants, vii, viii, ix, 1, 3, 6, 12, 18, 19, 24, 31, 34, 38, 47, 48, 52, 53, 55, 56, 58, 59, 60, 65, 66, 69, 70, 71, 73, 75, 76, 81, 92, 94, 95, 98, 99, 102, 103, 104, 105, 110, 111, 112, 113, 114, 115, 116, 119, 123

infection, 17, 18, 23, 38, 54, 111, 112

infectious agents, 3

inflammation, 40, 115

inflammatory disease, 24

inflammatory responses, 36, 40

inhibition, viii, 2, 35, 38, 40, 118

insulin, 4, 6, 24, 117, 118, 124

insulin sensitivity, 118

intensive care unit, 113

intestine, 21, 24, 32, 116

iron, 10, 14, 19, 20, 38, 47, 49, 51, 54, 66

L

lactation, v, ix, 3, 6, 12, 13, 14, 15, 19, 20, 24, 25, 32, 47, 49, 50, 57, 62, 70, 71, 73, 75, 77, 79, 80, 82, 84, 85, 87, 91, 92, 93, 94, 95, 96, 99, 100, 101, 102, 103, 105, 107, 109, 110, 111, 114, 115, 117, 118, 119, 120, 121, 123, 125

lactic acid, vii, viii, 2, 5, 31, 33, 43, 48, 49, 54, 62, 63, 65, 66, 67

Lactobacillus, 3, 21, 25, 26, 27, 28, 29, 31, 33, 35, 38, 39, 41, 42, 45, 46, 49, 50, 54, 55, 59, 60, 61, 63, 64, 65, 66, 67, 115

lactoferrin, 3, 12, 14, 17, 22, 38, 45, 47, 49, 54, 55, 66, 111, 112, 116, 117, 118

lactose, 7, 14, 16, 31, 36

leptin, 24, 33, 100, 117, 124, 125

linoleic acid, 15, 72, 83, 96, 114

lipids, 4, 10, 14, 40, 70, 81, 94, 96, 98, 99, 100, 102, 114, 116, 122

lysozyme, 4, 22, 55, 57, 116

M

macronutrients, 3, 96, 104, 116

maternal diet, viii, 32, 45, 58, 69, 71, 85, 86, 87, 90, 92, 93, 115

mature milk, ix, 2, 13, 14, 15, 32, 53, 70, 77, 78, 79, 81, 82, 83, 84, 85, 86, 88, 94, 95, 96, 97, 101, 103, 111, 118, 125

meta-analysis, 93, 98, 121, 122

metabolism, ix, 13, 24, 72, 101, 110

metabolism of fatty acids, 72, 74

metabolites, 37, 38, 73, 75, 80, 85
microbiota, 3, 6, 11, 27, 28, 32, 36, 45, 48, 53, 54, 56, 57, 58, 60, 62, 63, 66, 113, 115
micronutrients, 3, 15, 116
microorganisms, 14, 44, 111, 116
molecular mass, 44, 58
molecular structure, 20
molecular weight, 37
monounsaturated fatty acids, 90, 91, 97, 116

N

neonates, vii, viii, 1, 14, 20, 24, 29, 47, 54, 56, 69, 117, 119, 124
nervous system, 112
neurodevelopment, viii, 69, 75, 76, 81, 92
nutrients, 2, 6, 7, 13, 21, 34, 36, 40, 47, 53, 55, 67, 70, 71, 93, 95, 97, 98, 99, 100, 102, 105, 107, 110, 113, 114, 120, 121, 122, 125
nutrition, vii, viii, ix, 1, 3, 13, 32, 45, 48, 63, 69, 70, 72, 75, 76, 100, 102, 110, 111, 119
nutritional status, 32, 57, 88

O

obesity, ix, 33, 110, 112, 117, 124, 125
organism, 6, 44, 72, 73
organs, 13, 24, 42, 117
oxidation, 23, 73, 96, 118

P

pathogens, vii, ix, 2, 3, 9, 11, 14, 20, 36, 38, 55, 66, 110, 111
peptides, 3, 18, 21, 24, 37, 39, 55, 71, 112, 116, 119, 124
perinatal, 33, 45, 60, 65, 73, 75, 76

phosphatidylethanolamine, 40, 76, 118
phospholipids, 71, 99, 103, 104, 114, 118
polymerase chain reaction, 26, 60
polymorphism, 59
polyunsaturated fat, viii, 34, 61, 62, 69, 72, 77, 90, 92, 95, 96, 97, 100, 101, 103, 105, 111, 114, 118
polyunsaturated fatty acids, viii, 34, 62, 69, 72, 77, 90, 92, 95, 96, 97, 100, 101, 103, 111, 114, 118
positive correlation, 74, 86, 87, 117
pregnancy, 6, 13, 34, 51, 58, 61, 62, 63, 73, 81, 87, 94, 100, 101, 104
premature infant, 21, 81, 95, 98, 107, 112
preterm, viii, 33, 66, 69, 70, 80, 81, 82, 83, 84, 85, 92, 93, 94, 97, 98, 99, 101, 102, 103, 104, 106, 107, 111, 112, 113, 117, 120, 121, 122, 124
preterm delivery, 97, 98, 122
preterm infants, viii, 69, 81, 94, 106, 112, 113
prevention, ix, 6, 32, 45, 109, 110, 112, 119
principal component analysis, 88
probiotic, viii, 2, 5, 25, 28, 32, 35, 37, 44, 45, 49, 55, 59, 60, 61, 62, 113
probiotics, 28, 34, 35, 38, 45, 59, 63, 68
pro-inflammatory, 21, 23, 61
protection, ix, 2, 9, 11, 17, 19, 21, 44, 110, 111, 117, 119, 124
protein structure, 39
protein synthesis, 7
proteins, 3, 12, 14, 16, 17, 18, 19, 20, 25, 37, 38, 40, 46, 47, 50, 53, 57, 58, 60, 65, 75, 111, 113, 115, 116, 117, 124

R

receptors, 9, 21, 23, 113, 117
recommendations, iv, 46, 100, 101
requirements, 26, 52, 84, 95, 113

risk, 21, 45, 53, 55, 63, 65, 112, 113, 121, 125

S

saturated fat, 88, 90, 91, 97, 114
saturated fatty acids, 90, 97
small intestine, 10, 20, 52
species, 25, 36, 56, 60, 67
supplementation, 45, 57, 62, 73, 75, 76, 95, 99, 106, 122
synthesis, 12, 19, 40, 76, 81

T

term, viii, ix, 3, 11, 15, 33, 58, 66, 69, 70, 71, 77, 80, 82, 83, 84, 85, 92, 93, 94, 97, 98, 99, 101, 102, 103, 104, 106, 107, 110, 111, 113, 114, 115, 117, 118, 119, 120, 121, 122, 124
term milk, 77, 80, 82, 83, 84, 101, 111, 114
thickening agents, 44
transforming growth factor, 4, 23
transitional, 14, 15, 70, 77, 78, 79, 82, 83, 84, 85, 86, 89, 94, 101, 103, 118, 125
transitional milk, 14, 15, 70, 78, 79, 82, 83, 86, 94, 118
treatment, 6, 28, 41, 45, 49, 50, 57, 117, 124
tumor necrosis factor, 23, 114

U

upper respiratory tract, 59
urinary tract, 10, 112, 120
urinary tract infection, 112, 120

V

variations, 13, 16, 25, 26, 95
vascular endothelial growth factor (VEGF), 117
viruses, 20, 38, 40, 113, 116
visual acuity, 76
vitamin B1, 10, 17, 48
vitamin B12, 10, 17, 48
vitamins, 3, 14, 15, 21, 23

W

World Health Organization, 35, 107, 111, 119, 121
worldwide, 77, 84, 91, 98

Z

zinc, 14, 19, 47

A

α-linolenic acid, 72, 83